Lecture Notes in Mathematics

Edited by A. Dold and B. Eckmann

472

Probability-Winter School

Proceedings of the Fourth Winter School on Probability Held at Karpacz, Poland, January 1975

Edited by Z. Ciesielski, K. Urbanik, and W. A. Woyczyński

Springer-Verlag
Berlin · Heidelberg · New York 1975

Editors
Prof. Z. Ciesielski
Institute of Mathematics
Polish Academy of Sciences
Gdańsk Division
Abrahama 18
81825 Sopot
Poland

Prof. K. Urbanik and
Prof. W. A. Woyczyński
Institute of Mathematics
Wroclaw University
Grunwaldzki 2/4
50384 Wroclaw
Poland

AMS Subject Classifications (1970): 28A40, 28A45, 28A65, 60BXX, 60FXX, 60GXX, 60JXX, 60K35

ISBN 3-540-07190-3 Springer-Verlag Berlin · Heidelberg · New York
ISBN 0-387-07190-3 Springer-Verlag New York · Heidelberg · Berlin

Printed in Germany
Offsetdruck: Julius Beltz, Hemsbach/Bergstr.

FOREWORD FROM THE EDITORS

In the present volume we collect the written and often expanded versions of lectures delivered at the Fourth Winter School on Probability held on January 9-19,1975 in the Sudeten Mountains ski resort of Karpacz , Lower Silesia, Poland, and sponsored jointly by Wrocław University and the Institute of Mathematics of the Polish Academy of Sciences. This year the main topics of the School were: non-commutative probability theory(sometimes also called the geometry of quantum mechanics),random fields, and probabilities on linear spaces.Some of the contributions are of survey and/or expository character, and the others offer mainly an original material.

Taking an opportunity we would like to thank all who were helpful while these proceedings were being prepared : the Authors,Springer-Verlag and,in particular, Miss Teresa Bochynek who did the fine typing job.

Z.Ciesielski,
K.Urbanik,
W.A.Woyczyński

CONTENTS

WINTER SCHOOL
ON PROBABILITY
Karpacz 1975
Springer's LNM

472

STOCHASTIC SYSTEMS OF PARTICLES

By Z. Ciesielski
Institute of Mathematics, Polish Academy of Sciences

CONTENTS

0. Introduction

The aim of the first part of the lecture is the presentation of the passage to the thermodynamic limit in a model of the ideal gas and the proof of the existence of the state of thermodynamic equilibrium. In the second part we analyse the behaviour of the equilibrium state taking into account the stochastic dynamics. In the last part we talk about the deterministic dynamics in the one-dimensional case. Here the Wiener and Ornstein-Uhlenbeck processes come out in a natural way. These processes play an inportant role in the lecture as a whole. From the point of view of the Gibbsian mechanics the theory of Brownian motion developed by Einstein and Smoluchowski becomes more understandable. By no means the results presented in this lecture are claimed to be original, they can be found in the literature quoted at the end.

1. Classical mechanics.

We shall consider the system of N particles (material points) in $\underline{R} = R^3$. The state of such a system may be identified with the pair of vectors $(\overline{x},\overline{v}) \in \underline{R}^N \times \underline{R}^N$, where $\overline{x} = (x_1,\dots,x_N)$ is the position vector and $\overline{v} = (v_1,\dots,v_n)$ is the velocity vector; x_i, v_i and m_i stand, respectively, for the vectors of position, velocity and mass of i-th particle. Every such a system is described by its Hamiltonian

$$(1.1) \qquad H(\overline{x},\overline{v}) = \frac{1}{2}\sum_{i=1}^{N} m_i v_i^2 + U(\overline{x}) + U_0(\overline{x}) ,$$

where U is the potential of the interaction between particles and U_0 is a potential of outside forces. In the case of a system in which the particles interact only pairwise, U is of the form

$$(1.2) \qquad U(\overline{x}) = \sum_{i<j} \Phi(|x_i - x_j|) ,$$

and the graph of a typical function Φ looks as follows

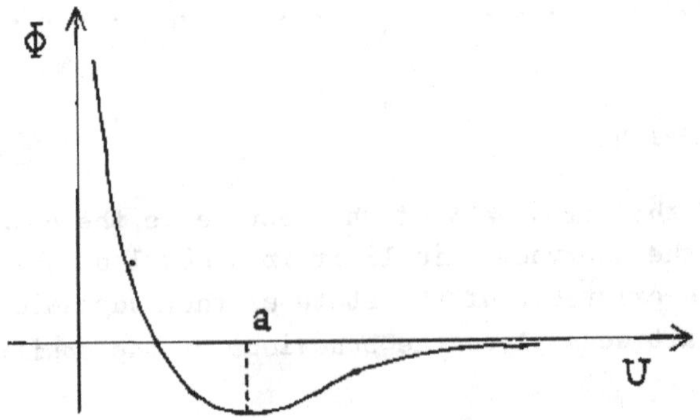

where $(0,a)$ corresponds to the domain of repulsion and (a,∞) to the domain of attraction. The potentials U, U_0 may, on some subsets, take the value $+\infty$.

The model given above, where $U = 0$, corresponds to the ideal gas. The model of the gas cosisting of hard balls (of the same size) may be obtained putting $\Phi(u) = +\infty$ for $0 < u < a$ and $\Phi(u) = 0$ for $u \geqslant a$.

The fact that our system of particles is contained in the box $V \subset \underline{R}$ (bounded domain) means that the potential of outside forces is of the form

$$(1.3) \qquad U_0(\overline{x}) = \begin{cases} +\infty & \text{for } \overline{x} \notin V^N , \\ 0 & \text{for } \overline{x} \in V^N \end{cases}$$

In what follows we shall assume that the potential U_0 is of the form (1.3). In this case, the space of admissible states of our system reduces to $S = V^N \times \underline{R}^N$.

2. Statistical mechanics

In the real system, in which the number of particles is of order 10^{23}, conducting precise observations of the positions and velocities of particular partcles or even the exact measurement of other physical quantities such as energy entropy, pressure, turns out impossible. So there is real necessity of statistical approach to such large systems of particles. The notion of the state as a point in the space S fails to be adequate. It has to be modified and this is the point where the statistical mechanics begins. Instead of talking about states as points $(\overline{x},\overline{v}) \in S$ we may think about them as probability distributions which may only say what is the probability that the state of the system is in a given set $A \subset S$. Each state $(\overline{x},\overline{v})$ of the classical mechanics may be here identified with the probabilistic measure concentrated at this point

$$(2.1) \qquad \delta_{(\overline{x},\overline{v})}(A) = \begin{cases} 1 & \text{if } (\overline{x},\overline{v}) \in A , \\ 0 & \text{if } (\overline{x},\overline{v}) \notin A . \end{cases}$$

However, in this way, if we took the set of all probability measures on S as the set of "states", we would get to big collection of them. It would contain the classical states what would cause troubles again. So we shall take as the space of states a certain subset of the set of all probability measures on S . There is in S the natural algebra of Borel subset and the Lebesgue measure m defined on it. The <u>state space</u> of our system of N particles will be the set of probability Borel measures on S that are absolutely continuous with respect to m . The possibility of the exact description of $(\overline{x},\overline{v})$ is therefore excluded (the measure of the form (2.1) would give such a possibility). It is handy to identify the set of states with the set of their densities with respect to m i.e. with

$$\sigma = \left\{ f : f \geqslant 0,\ f \in L^1(S,m),\ \int_S f dm = 1 \right\} . \tag{2.2}$$

Borel functions on S will be called physical quantities. The Hamiltonian H given by formula (1.1) is an example of such a quantity. If the system is in the state $f \in \sigma$ then (S,ν_f) , where $d\nu_f = fdm$, is a probability space, and the physical quantities are random variables on it. Thus we may speak about their mean values in the state f e,g. the mean energy.

$$E(f) = \int_S H(\overline{x},\overline{v})\ f(\overline{x},\overline{v})\ dm . \tag{2.3}$$

In what follows we assume that the system is isolated. It implies that the mean energy E of the system is given and the equality $E = E(f)$ together with formula (2.3) describes all admissible states f i.e. those in which our isolated system may be found. If the density f of an admissible state is concentrated around $(\overline{x},\overline{v})$ then we have a lot of information about the system. The minimal information about the system is provided in the admissible state corresponding to the "most uniform" distribution or, in other words, in the state with maximal entropy. Here, entropy is the value of the functional

$$\varepsilon(f) = - \int_S f \log f dm , \tag{2.4}$$

It is a good measure of uncertainty of the state of the system. It is not hard to check that the admissible state of the form

(2.5) $$f_\beta(\bar{x},\bar{v}) = Z^{-1}(\beta)\exp\left[-\beta H(\bar{x},\bar{v})\right] ,$$

where

$$Z(\beta) = \int_S \exp(-\beta H)\, dm ,$$

has the maximal entropy, i.e.

$$(f_\beta) = \sup\left\{ (f) : E = E(f)\right\} ,$$

where $E = E(f_\beta)$. Under certain assumptions about the Hamiltonian H the correspondance $E \to \beta$ given by equation $E = E(f_\beta)$ is one-to-one. The state f_β , defined in such a way is called the equilibrium state or the Gibbs state. In the equilibrium state one can talk about temperature. It turns out that $\beta = (kT)^{-1}$, where k is the Boltzmann constant and T is the temperature.

Notice, that from the shape of Hamiltonian (1.1) and from formula (2.5) it follows that the vectors of position and velocity are stochastically independent. Taking into account condition (1.3) we get for the ideal gas consisting of identical particles (of mass $m_i = 1$) that

(2.6) $$f_\beta(\bar{x},\bar{v}) = h(\bar{x})\cdot g_\beta(\bar{v}) , \quad (\bar{x},\bar{v}) \in S ,$$

where $h(\bar{x}) = |V|^{-N}$ is the density of the uniform distribution on V^N , and g_β is the Gaussian density on $\underline{R}^N = R^{3N}$ with the mean zero and the covariance matrix $\beta^{-1}I$ i.e.

$$g_\beta(\bar{v}) = \left(\frac{\beta}{2\pi}\right)^{\frac{3N}{2}} \exp\left\{-\frac{\beta}{2}\sum_{i=1}^{N} v_i^2\right\} \quad \text{for } \bar{v} \in \underline{R}^N .$$

It is easily checked that in this case (of the ideal gas) we have for the mean energy and mean entropy in the state (2.6) the following formulas

(2.7) $$E = E(f_\beta) = \frac{3}{2}\cdot\frac{N}{\beta} = \frac{3}{2} NkT ,$$

$$\mathcal{E}(f_\beta) = N\,[\log |V| + \tfrac{3}{2}\log(\tfrac{2\pi e}{\beta})]\ . \tag{2.8}$$

In what follows we shall be interested in the behaviour of our model while passing to the thermodynamic limit i.e. while we simultaneously expand the box V and increase the number of particles N keeping the density ρ constant, i.e.

$$\frac{N}{|V|} \to \rho\ , \qquad V \uparrow \underline{R} \tag{2.9}$$

3. <u>Point measures as realizations of stochastic processes</u>

The aim of the present section is to discuss the mathematical apparatus which later on enables us to construct such a probabilistic model for finite and infinite system of particles of the ideal gas in which the passage to the thermodynamic limit will be interpreted as the convergence of finite-dimensional distributions of corresponding stochastic processes.

In what follows let Z_+ be the set of non-negative integers, $k \in Z_+$, $k \geq 1$ and let B^{k^+} be the σ - algebra of all Borel subsets of R^k. $\bar{Z}_+$ will stand for $Z_+ \cup \{+\infty\}$.

The <u>point measure</u> on R^k is a mapping $\xi : B^k \to \bar{Z}_+$ with the following properties

(3.1) $\qquad \xi(A) < \infty \qquad$ for relatively compact A ;

(3.2) $\qquad \xi(A \cup B) = \xi(A) + \xi(B) \qquad$ for $A \quad B = \emptyset$;

(3.3) If A_1 is bounded and $A_n \searrow \emptyset$, then $\xi(A_n) = 0$ for n large enough.

The set of all point measures on R^k will be denoted by $\underline{M}$.

THEOREM 3.1. $\xi \in \underline{M}$ <u>if and only if for relatively compact</u> $A \in B^k$

$$\xi(A) = \sum_{x_i \in A} n_i\, I_A(x_i)\ , \quad n_i \neq 0 \tag{3.4}$$

where $\{x_i\}$ is a countable subset which does not contain the (finite) cluster point and $\{n_i\} \subset Z_+$. Moreover, representation (3.4) is unique up to the numbering.

We omit the easy proof of this theorem.

In the space $\underline{M}$ we distinguish the σ - algebra $\mathfrak{M}$ of subsets generated by the sets of the form

$$\left\{\xi \in \underline{M} : \xi(G) < a\right\},$$

where G is an open and bounded subset of R^k and a is a positive real number.

Each probability measure $P : \mathfrak{M} \to [0,1]$ defines a stochastic process $\{\xi(A), A \in B^k\}$ on the probability space $(\underline{M}, \mathfrak{M}, P)$. We write down the necessary conditions for its finite- dimensional distributions

(3,5) $$Q(A_1,\dots,A_n; r_1,\dots,r_n) = P\left\{\xi(A_i) = r_i,\ i = 1,\dots,n\right\},$$

where $(r_1,\dots,r_n) \in Z_+^n$, $A_i \in B^k$, $i = 1,\dots,n$ and A_i are relatively compact.

(3.6) $$Q(A_1,\dots,A_n; r_1,\dots,r_n) = Q(A_{\pi(1)},\dots,A_{\pi(n)}; r_{\pi(1)},\dots,r_{\pi(n)})$$

where π is an arbitrary permutation of the numbers $1,2,\dots,n$.

(3.7) $Q(A_1,\dots,A_n; \cdot)$ is a probability distribution on Z_+^n ;

(3.8) $$\sum_{r_{n+1}=0}^{\infty} Q(A_1,\dots,A_{n+1}; r_1,\dots,r_{n+1}) = Q(A_1,\dots,A_n; r_1,\dots,r_n)$$

(3.9) If $A = \bigcup_{i=1}^{n} A_i$, $A_i \cap A_j = \emptyset$ for $i \neq j$, then

$$Q(A_1,\dots,A_n,A; r_1,\dots,r_n,r)$$

$$= \begin{cases} 0 & \text{if } r \neq r_1 + \dots + r_n , \\ Q(A_1,\dots,A_m; r ,\dots,r_m) & \text{if } r = r_1 + \dots + r_n . \end{cases}$$

(3,10) If $A_n \searrow \emptyset$, then $Q(A_n; r) \searrow 0$ for each $r \in Z_+$.

THEOREM 3.2. (T.E.Harris). If the family $\{Q(A_1,\dots,A_n;r_1,\dots,r_n)$ where $A_i \in B^k$ are relatively compact, $(r_1,\dots,r_n) \in Z_+^n,\ n \geq 1\}$ satisfies conditions (3.6) - (3.10) then there is unique probability measure P on $(\underline{M},\mathfrak{M})$ such that (3.5) is satisfied.

We omit the proof (cf. [16], p. 88, th 3.1 and [5], th.(2.3), p. 88) .

Now, let us take a short look at conditions (3.6) - (3.10). Checking them for a given family $\{Q\}$ may be troublesome. Assume, that $Q(A_1,\dots,A_n;\ r_1,\dots,r_n)$ is defined only for $(A_1,\dots,A_n)$ for which $A_i \cap A_j = \emptyset$, $i \neq j$. Assume furthermore that $Q(.)$ with such a domain satisfies (3.6), (3.7), (3.8), (3.10) but instead of (3.9) it satisfies condition

$$(3.11) \qquad Q(A_1,\dots,A_n;\ r_1,\dots,r_n)$$

$$= \Sigma\ Q(A_{1,1},\dots,A_{1k_1},\dots,A_{n1},\dots,A_{nk_n};r_{11},\dots,r_{1k_1},\dots,r_{n1},\dots,r_{nk_n})$$

which is satisfied for $A_i \cap A_j = \emptyset$, $i \neq j$ and for arbitrary partitions

$$A_i = \bigcup_{j=1}^{k_i} A_{ij}\ , \qquad A_{ij} \cap A_{ih} = \emptyset \quad \text{for} \quad j \neq h\ ,$$

where the summation in (3.11) ranges over all the (r_{ij})'s such that

$$r_i = r_{i1} + \dots + r_{ik_i} \qquad \text{for} \quad i = 1,\dots,r\ .$$

Then, $Q(.)$ may be extended in such a (unique) way that conditions (3.6) - (3.10) are satisfied for any n-tuples $(A_1,\dots,A_n)$ of Borel relatively compact sets A_j . So we arrive at the conclusion that Theorem 3.2 and the above remarks give the following corollary. If Q is defined for pairwise disjoint A_i and satisfies conditions (3.6), (3.7), (3.8), (3.11) and (3.10), then there is a probability measure on $(\underline{M},\mathfrak{M})$ for which formula (3.5) holds.

4. Probabilistic model of the ideal gas in the equilibrium state. Thermodynamic limit.

Let us return to the system of N particles of the ideal gas contained in the box V. In the equilibrium state the probability density of the vector $(\overline{x},\overline{v}) \in S = V^N \times \underline{R}^N$ is given by the formula (2.6). It means that the vectors $(x_1,v_1),\dots,(x_N,v_N)$ are independent and identically distributed in $V \times \underline{R}$ with the distribution F_V, where $dF_V = f_V dm$, m is the Lebesgue measure on $V \times \underline{R}$ and

$$(4.1) \qquad f_V(x,v) = \frac{1}{|V|}\left(\frac{\beta}{2\pi}\right)^{\frac{3}{2}} \exp\left(-\frac{\beta}{2}v^2\right).$$

Now, for fixed parameters N and V we construct a stochastic process $(\underline{M},\mathfrak{M},P_{N,V})$ which is a probabilistic model for the above system of N particles of theideal gas contained in the box V in the equilibrium state. For $n \geq 1$, $(r_1,\dots,r_n) \in Z_+^n$, $A_i \in B^6$ for $i = 1,\dots,n$; $A_i \cap A_j = \emptyset$, $i \neq j$, we define

$$(4.2) \qquad Q_{N,V}(A_1,\dots,A_n;\, r_1,\dots,r_n)$$

$$= \frac{N!}{r_1!\,\dots\,r_n!\,(N-r_1-\dots-r_n)!}\, p_1^{r_1}\dots p_n^{r_n}(1-p_1-\dots-p_n)^{N-r_1\dots-r_n}$$

where $p_i = F_V[A_i \cap (V \times \underline{R})]$. Let $F_{N,V} = F_V \otimes \dots \otimes F_V$ (N-times). Evidently, $dF_{N,V} = f_\beta dm$, where f_β is given by formula (2.6), and m denotes the Lebesgue measure on S. In the language of the probability space $(S,F_{N,V})$, $Q_{N,V}$ may be expressed as follows

$$(4.3) \qquad Q_{N,V}(A_1,\dots,A_n;\, r_1,\dots,r_n)$$

$$= F_{N,V}\left[\sum_{i=1}^{N} I_{A_s}(x_i,v_i) = r_s\,,\quad s = 1,\dots,n\right].$$

It is not difficult to check that $Q_{N,V}(\cdot)$, defined by formula (4.2) for $A_i \cap A_j = \emptyset$, $i \neq j$, satisfy conditions (3.6), (3.7), (3.8), (3.11) and (3.10). Consequently, in view of what was said in Section 3, there is probability measure $P_{N,V}$ on $(\underline{M},\mathfrak{M})$ such that for

$A_i \cap A_j = \emptyset$, $i \neq j$,

$$(4.4) \qquad P_{N,V}[\xi(A_s) = r_s \ , \ s = 1,\ldots,n] = Q_{N,V}(A_1,\ldots,A_n;r_1,\ldots,r_n)$$

So that $(\underline{M},\mathfrak{M},P_{N,V})$ is probabilistic model we looked for. According to formulas (4.3) and (4.4), and by uniqueness of the measure $P_{N,V}$, it follows that $P_{N,V}$ is concentrated on the set of those $\xi \in \underline{M}$ for which in theorem 3.1, applied to the point measures on $R^k = \underline{R} \times \underline{R}$, $n_i = 1$ holds for all i .

Now, we are ready to perform the passage to the thermodynamic limit. We expand the box and increase the number of particles in such a way that condition (2.9) is satisfied. Notice, that for Borel and relatively compact $A \subset \underline{R} \times \underline{R} = R^6$, for big V

$$F_V(A \cap V \times \underline{R}) = \frac{1}{|V|} \lambda_\beta(A) \quad ,$$

Where $\lambda_\beta = m \otimes \nu_\beta$, m is the Lebesgue measure on $\underline{R}$ and

$$\nu_\beta(dv) = (\frac{\beta}{2\pi})^{\frac{3}{2}} \exp\,(-\,\frac{\beta}{2}\,v^2)\ dm \ , \quad v \in \underline{R} \ .$$

For given relatively compact $A_i \subset \underline{R}^2$, $i = 1,\ldots,n$ we get that if $\frac{N}{|V|} \to \rho$ then

$$Np_i \to \rho\ \lambda_\beta\ (A_i) \qquad \text{for} \quad i = 1,\ldots,n \ .$$

By elementary computation we get from (4.2) and (4.5) that if $\frac{N}{|V|} \to \rho$, $A_i \cap A_j = \emptyset$, $i \neq j$, then

$$(4.6) \qquad Q_{N,V}(A_1,\ldots,A_n;\ r_1,\ldots,r_n) \to Q_\rho(A_1,\ldots,A_n;\ r_1,\ldots,r_n) \ ,$$

where

$$(4.7) \qquad Q_\rho(A_1,\ldots,A_n;\ r_1,\ldots,r_n) = Q_\rho(A_1,r_1)\ldots\ Q_\rho(A_n,r_n)$$

and

(4.8) $$Q_\rho(A;\ r) = e^{-\rho\lambda_\beta(A)}\ \frac{[\rho\lambda_\beta(A)]^r}{r!}$$

Also for $Q_\rho(.)$ conditions (3.6), (3.7), (3.8), (3.11), (3.10) are easy to check for disjoint A_i. Theorem 3.2 and following it corollary imply existence of the probability measure P_ρ on $(\underline{M},\mathfrak{M})$ such that for $A_i \cap A_j = \emptyset$, $i \neq j$,

$$P_\rho[\xi(A_i) = r_i,\ i = 1,\dots,n] = \prod_{i=1}^{n} Q_\rho(A_i,r_i)\ ,$$

where $Q_\rho(A,r)$ is given by formula (4.8).

Extension to the sets A_i that are not necessarily disjoint is related to the convergence of $P_{N,V}$ to P_ρ in finite-dimensional distributions

(4.9) $$P_{N,V}(\xi(A_i) = r_i,\ i=1,\dots,n) \to P_\rho(\xi(A_i) = r_i,\ i=1,\dots,n)$$

as $\frac{N}{|V|} \to \rho$ for any n, $(r_1,\dots,r_n) \in Z_+^N$ and Borel and relatively compact A_i. The space $(\underline{M},\mathfrak{M},P_\rho)$ is the probabilistic model we looked for describing the infinite system of particles of the ideal gas with the density ρ in the equilibrium state. It is worth mentioning that P_ρ is concentrated on the set of those ξ for which $n_i = 1$ for all i (cf. th.3.1). This fact follows immediately from formulas (4.7), (4.8) and the th. 3.1 (cf. Doob [2]). So we see that the equilibrium state of an infinite system of particles of the ideal gas with the density ρ may be identified with the Poisson random measure ξ with expectation $E_\rho(\xi(A)) = \rho\lambda_\beta(A)$.

5. Invariance of the equilibrium state.

In the sequel, by the equilibrium state ξ of an infinite system of particles of the ideal gas with density ρ over the probability space $(\Omega,\mathfrak{F},P)$ we shall mean the Poissonian random measure ξ such that

(5.1) $$E(d\xi) = \rho\ d\ \lambda_\beta$$

$$(5.2) \qquad P\left\{\xi(.) \in \underline{M}\right\} = 1 \quad .$$

The observed in the domain $\bar{V} = V \times \underline{R} \subset \underline{R}^2$ subsystem of N particles of the ideal gas in the equilibrium state ξ is described by conditional probabilities

$$(5.3) \qquad P\left\{\xi(A_i) = r_i,\ i = 1,\dots,n \mid \xi(\bar{V}) = N\right\}$$

$$= Q_{N,V}(A_1,\dots,A_n;\ r_1,\dots,r_n) \ , \qquad A_i \cap A_j = \emptyset \ \text{ for } \ i \neq j \ .$$

From the considerations of the preceding sections it follows that the finite-dimensional distributions of the random measure ξ are determined by the conditional probability given by formula (5.3). On the other hand (5.3) corresponds to the system of N independent random vectors $z_i = (x_i, v_i)$ in $\bar{V}$ with the distribution F_V of each of them and with satisfied.(4.3).

Let $\{z_i(t),\ 0 \leq t < \infty\}$, $i = 1,\dots,N$ denote N independent copies of the Markov process in $\underline{R}^2$, with given stationary transition probability $(P(t,z,A)$, $A \subset \underline{R}^2$. All the observed particles z_i , $i = 1,\dots,N$, are subject to independent Markovian motions with the same transition probability $P(t,z,A)$ and with $P\{z_i(0) = z_i\} = 1$. Consequently, for $A_i \cap A_j = \emptyset$, $i \neq j$, after the time $t > 0$, the distribution of $z_i(t)$, $i = 1,\dots,N$, is given by the formula

$$Q^t_{N,V}(A_1,\dots,A_n;\ r_1,\dots,r_n)$$

$$= \frac{N!}{r_1!\dots r_n!(N-r_1-\dots-r_n)!}\, p_1^{r_1}(t)\dots p_n^{r_n}(t)[1-p_1(t)-\dots-p_n(t)]^{N-r_1-\dots-r_n},$$

where

$$p_i(t) = \int_{\bar{V}} P(t,z,A_i \cap \bar{V})dF_V(z) \quad .$$

From these remarks it follows that for fixed $t > 0$ the family $\{Q^t_{N,V}\}$ ensures the existence of the Poissonian measure ξ_t such that

$$(5.4) \qquad E(d\,\xi_t) = \rho\ d\ \lambda^t_\beta \quad ,$$

(5.5) $$P\left\{\xi_t(.) \in \underline{M}\right\} = 1 ,$$

(5.6) $$\lambda_\beta^t(A) = \int_{\underline{R}^2} P(t,z,A)\, \lambda_\beta(dz) , \quad \lambda_\beta = m \otimes \nu_\beta \quad .$$

Moreover,

$$P\left\{\xi_t(A_i) = r_i,\ i = 1,\dots,n \mid \xi_t(\bar{V}) = N\right\} = Q^t_{N,V}(A_1,\dots,A_n; r_1,\dots,r_n) .$$

THEOREM 5.1. (Doob [2], p. 406). <u>Let for the equilibrium state</u> ξ <u>of the ideal gas and for given</u> $t > 0$ <u>the random Poisson measure</u> ξ_t <u>be defined as above. Then</u> ξ_t <u>corresponds to equilibrium state of the ideal gas with density</u> ρ <u>iff</u> $\lambda_\beta^t = \lambda_\beta$ <u>or else iff</u> λ_β <u>is invariant under the action of</u> $P(t,z,A)$ <u>i.e. iff</u>

(5.8) $$\lambda_\beta(A) = \int_{\underline{R}^2} P(t,z,A)\, \lambda_\beta(dz), \quad A \in B^6 .$$

The proof is straightforward. It is sufficient to compare conditions (5.1), (5.2) with (5.4), (5.5), and apply formula (5.6).

One may relax the assumption that at the moment $t = 0$ the gas is in the equilibrium state and still get that after long enaugh time t , ξ_t will be close to the equilibrium state. This kind of phenomenon is described by results of Stone and Dobrushin but it will not be discussed here.

6. Brownian motion.

We shall talk here about the stochastic process which is a mathematical model of what is known as <u>Brownian molecular motion</u>. To get the first feeling of this physical phenomenon we quote only the title of the first article by Einstein on this subject: "On the movement of small particles suspended in a stationary liquid demanded by molecular kinetic theory of heat", (cf. [3], p.1). The physical theory of Brownian motion was given by Einstein and Smoluchowski, and basing on their theory Wiener constructed precise mathematical model which today is called the Wiener process.

The relation between mathematics and physics is expressed by the fact that the Wiener prosess is the homogeneous Markov process in $\underline{R}$ with the density of the transition probability $p(t,x,y)$ being the fundamental solution of the diffusion equation

$$\frac{\partial u}{\partial t} = D \Delta u ,$$

where D is the diffusion constant and Δ is the Laplace operator in $\underline{R} = R^3$. From the above it follows that

$$p_3(t,x,y) = \frac{1}{(4\pi Dt)^{3/2}} \exp\left(- \frac{|x-y|^2}{4Dt}\right) \tag{6.1}$$

It is easy to check (either using the characteristic functions or directly) that the following Smoluchowski's equation is satisfied

$$p(t+s,x,y) = \int_{\underline{R}} p(s,x,z)\, p(t,z,y)\, dz . \tag{6.2}$$

Now, let $C_k = C(R_+,R^k)$ be the space of continuous functions on R_+ with values in R^k , and let $\mathcal{F}_k$ be the σ - algebra of Borel subsets of C_k . Using formula (6.1) and equation (6.2) one shows that for each $x \in R$ there is a probability space $(C_3,\mathcal{F}_3,P_x)$ with the folfownig properties:

1. $P_x\{x(0) = x\} = 1$;
2. $\{x(.),\ t \in R_+\}$ is a Markov process over $(C_3,\mathcal{F}_3,P_x)$;
3. For any $A \in B^3$ and for any $t > s \geqslant 0$ with probability P_x equal to 1

$$P_x\left\{x(t) \in A \mid x(s)\right\} = \int_A p(t-s,x(s),z)dz .$$

From the very construction it follows that the realizations of the process $(C_3,\mathcal{F}_3,P_x)$ are continuous and this process is called the three-dimensional <u>Brownian motion</u> starting at $x \in \underline{R}$. The <u>Wiener process</u> is the same process but starting at zero i.e. $(C_3,\mathcal{F}_3,P_0)$. As a probability space $(C_3,\mathcal{F}_3,P_0)$ is called the <u>Wiener space</u>.

If $x = (x^1,x^2,x^3)$ and $y = (y^1,y^2,y^3)$ then the formula (6.1) may be written as follows

$$p_3(t,x,y) = p(t,x^1,y^1)\, p(t,x^2,y^2)\, p(t,x^3,y^3) , \tag{6.3}$$

where

$$p(t,u,v) = \frac{1}{\sqrt{4\pi Dt}} \exp\left(-\frac{1}{4Dt}(u-v)^2\right) .$$

We infer from formula (6.3) that the Brownian motion $(C_3, \mathcal{F}_3, P_x)$ i.e. the process $x(t) = (x^1(t), x^2(t), x^3(t))$, has as the components independent Markov processes with continuous realizations and enjoying the property that for each $u \in R$ there is a probability space $(C_1, \ _1, P_u)$ such that

1. $P_u\{x^i(0) = u\} = 1$,
2. $\{x^i(t),\ t \in R_+\}$ is the Markov process over $(C_1, \mathcal{F}_1, P_u)$,
3. For $E \in B^1$ and for $t > s \geq 0$ with the probability P_u equal to 1

$$P_u\left\{x^i(t) \in A \mid x^i(s)\right\} = \int_E p(t-s, x(s), z)dz$$

Notice, that $C_3 = C_1 \times C_1 \times C_1$, $\mathcal{F}_3 = \mathcal{F}_1 \otimes \mathcal{F}_1 \otimes \mathcal{F}_1$ and for $x = (x^1, x^2, x^3)$, $P_x = P_{x^1} \otimes P_{x^2} \otimes P_{x^3}$. So, the investigation of the Brownian motion $(C_3, \mathcal{F}_3, P_x)$ can be reduced to the investigation of its independent components $(C_1, \mathcal{F}_1, P_u)$ being the one-dimensional Brownian motions.

The Brownian motion in R^k is defined as the probability space $(C_k, \mathcal{F}_k, P_z)$ being the k-fold product of space $(C_1, \mathcal{F}_1, P_{z^i})$, where $z = (z^1, \ldots, z^k)$, and the <u>Wiener process with independent components</u> is $(C_k, \mathcal{F}_k, P_0)$. The realizations of the Brownian motion will be denoted by $w(t) = (w^1(t), \ldots, w^k(t))$. For each $i = 1, \ldots, k$, $w^i(t)$ is the process with the following properties

1^0 $w^i(t)$ is Gaussian ;
2^0 $E(w^i(t) - w^i(s)) = 0$;
3^0 $w^i(t)$ has independent increments ;
4^0 $E(w^i(t) - w^i(s))^2 = 2D\,|t-s|$

Notice that the independence of components may be included in the condition

$$4^0_a.\quad E(w^i(t) - w^i(s))(w^j(t) - w^j(s)) = 2D\delta_{ij}|t-s| .$$

Changing the system of coordinates it is not hard to construct the process $w(t)$ which satisfies the conditions 1^0, 2^0, 3^0 and

4^o_b. $E(w^i(t) - w^i(s))(w^j(t) - w^j(s)) = c^{ij}|t-s|$,

where $\underline{C} = (c^{ij})$, $i,j = 1,\ldots,k$ is a positive-definite matrix and is called the <u>covariance</u> <u>matrix</u>.

7. <u>Stochastic</u> <u>integral</u> <u>of</u> <u>Paley-Wiener-Zygmund</u>.

Let $w(t)$ be the one-dimensional Wiener process with variance σ^2 and let $0 = t_o < t_1 < \ldots < t_n < \infty$, $I_i = [t_{i-1}, t_i)$,

$$f(t) = \sum_{i=1} c_i \chi_i(t) , \qquad \chi_i(t) = \begin{cases} 1 & \text{for} \quad t \in I_i \\ 0 & \text{for} \quad t \overline{\in} I_i . \end{cases}$$

It is easy to check that

$$E(Uf)^2 = \sigma^2 \int_0^\infty f^2(t)dt \quad , \tag{7.1}$$

where

$$Uf = \frac{1}{\sigma} \int_0^\infty f\, dw \tag{7.2}$$

is the Stieltjes integral. We may proceed in this way because w has continuous realizations.

The formula (7.1) permits to extend U , given by the formula (7.2), to the unique isometric embedding

$$U : L^2(0,\infty) \to L^2 (C_1, \mathcal{F}_1, P_o) .$$

So as the definition of the PWZ integral we take

$$\int_0^\infty f\, dw \overset{df}{=} \sigma\, Uf \qquad \text{for each} \quad f \in L^2(0,\infty) .$$

and

$$\int_0^t f\,dw \overset{df}{=} \int_0^\infty f\cdot\chi_t\,dw \quad , \quad \text{where} \quad \chi_t(u) = \begin{cases} 1 \; , & u \leq t \; , \\ 0 \; , & u > t \; . \end{cases}$$

8. Stochastic equations.

We restrict our attention to the stochastic equations in which the PWZ integral appears only but not the more general Ito's integral. Like, in the theory of differential equations the theorems on existence and uniqueness may be proved for non-linear equations but the general solutions one knows how to find for linear equations only.

Let $w(t)$ be the Wiener process in R^k with the covariance matrix $C = (c^{ij})$. Then for the functions on R^k that are sufficiently smooth

$$\lim_{t\to 0_+} \frac{E(f(z+w(t))) - f(z)}{t} = \frac{1}{2} \sum_{i,j=1}^{k} c^{ij} \frac{\partial^2 f(z)}{\partial z^i \partial z^j}$$

and so the differential operator

$$A_0 = \frac{1}{2} \sum_{i,j=1}^{k} c^{ij} \frac{\partial^2}{\partial z^i \partial z^j} \tag{8.1}$$

is the infinitesimal generator for the Wiener process w .

In the further considerations the positive definiteness of C is not essential. It is sufficient to assume that C is non-negative definite. The proofs of the two following theorems may be found in the book by E.Nelson [10].

THEORÊM 8.1. *Let $b : R^k \to R^k$ satisfy the global Lipschitz condition on R^k . Then for each $z_0 \in R^k$ there exists exactly one process $z(t)$, $0 \leq t < \infty$, such that*

$$z(t) = z_0 + \int_0^t b(z(s))ds + w(t) - w(0) \; . \tag{8,2}$$

The process $z(t)$ has continuous realizations, and it is Markovian with the infinitesimal generator

$$Af = b\,.\,\nabla t + A_0 f\ , \quad f \in C_0^2(R^k)\ ,$$

<u>and</u> $C_0^2(R^k)$ <u>is contained in the domain of</u> A <u>and</u> A_0 <u>is given by formula</u> (8.1).

It is convenient to write equation (8.2) in the equivalent differential form:

$$dz(t) = b(z(t))dt + dw(t)$$

THEOREM 8.2. <u>Let</u> $\underline{A} : R^k \to R^k$ <u>be linear and let</u> $f \in C_k$. <u>Then the solution of the equation</u>

$$dz(t) = \underline{A}z(t)dt + f(t)dt + dw(t)\ , \quad z(0) = z_0$$

<u>for</u> $t > 0$ <u>is given by the formula</u>

$$z(t) = e^{\underline{A}t}\, z_0 + \int_0^t e^{\underline{A}(t-s)} f(s)ds + \int_0^t e^{\underline{A}(t-s)} dw(s)\ .$$

<u>The process</u> $z(t)$ <u>is Gaussian with mean</u>

$$E(z(t)) = e^{\underline{A}T}\, z_0 + \int_0^t e^{\underline{A}(t-s)} f(s)ds$$

<u>and the covariance</u>

$$r_{ij}(t,s) = E(z^i(t) - E(z^i(t)))(z^j(s) - E(z^j(s)))$$

<u>given by the formula</u>

$$(8.3) \qquad r(t,s) = \begin{cases} e^{\underline{A}(t-s)}\left(\int_0^s e^{\underline{A}r}\, \underline{C}\, e^{\underline{A}^T r} dr\right) & \text{for } t \geq s\ . \\ \left(\int_0^t e^{\underline{A}r}\, \underline{C}\, e^{\underline{A}^T r}\, dr\right) e^{\underline{A}(s-t)} & \text{for } t \leq s\ . \end{cases}$$

9. Ornstein - Uhlenbeck process.

The Wiener process as a model describing the position of a particle bombarded by atoms does not quite fit to reality as its realizations being continuous are everywhere non-differentiable with probability 1 . And it is quite hard to imagine the motion of a particle without the velocity.

So now we shall derive the stochastic equation of Langevin which as its solution has the Ornstein-Uhlenbeck process. This process provides a more realistic model of a particle moving under collisions with atoms.

Let $x(t)$ be the position of the particle in $\underline{R}$ and $v(t)$ its velocity at the moment t i.e.

$$v(t) = \frac{dx(t)}{dt} \quad . \tag{9.1}$$

On one hand the force acting on the particle is

$$F(t) = ma(t)$$

where a is the acceleration i.e.

$$a(t) = \frac{dv(t)}{dt}$$

On the other hand

$$F = F_0 + F_1 \,, \tag{9.2}$$

where F_0 is the force caused by friction which has the direction opposite to v i.e.

$$F_0(t) = -\gamma m v(t) \,,$$

γ = friction coefficient, and F_1 is the fluctuation force.

$$F_1(t) = m\,\frac{dw(t)}{dt} \,,$$

where $w(t)$ is the Wiener process in $\underline{R}$ with the covariance matrix

$\underline{C}$ chosen correspondingly.

Equations (9.1) and (9.2) may be written in the form of linear stochastic equation

$$(9.3) \qquad d\binom{x(t)}{v(t)} = \underline{A} \cdot \binom{x(t)}{v(t)} dt + d\binom{0}{w(t)}$$

In this equation $\underline{A}$ is a 6×6 - matrix of the form

$$A = \left(\begin{array}{c|c} 0 & I \\ \hline 0 & -\gamma I \end{array}\right)$$

and

$$\binom{0}{w(t)}$$

is the degenerated Brownian motion in $\underline{R}^2$, which is performed in the space of velocities only and has the following 6×6 covariance matrix

$$(9.4) \qquad \underline{C} = \left(\begin{array}{c|c} 0 & 0 \\ \hline 0 & \sigma_0^2 I \end{array}\right)$$

By the physical argument it is known that $\sigma_0^2 = 2\gamma^2 D$, where D is the diffusion constant. The matrix $\underline{C}$ is of course non-negative definite and so the process

$$z(t) = \binom{x(t)}{v(t)}$$

which is the solution of equation (9.3) is also the Markov process. It is called the Ornstein-Uhlenbeck process. From the theorem 8.2 it follows that it is Gaussian. From the shape of matrix (9.4) and equation (9.3) we infer that the process $z(t)$ has three independent components

$$z^i(t) = \binom{x^i(t)}{v^i(t)}$$

each being a solution of the equation

$$(9.5)\qquad d\begin{pmatrix} x^i(t) \\ v^i(t) \end{pmatrix} = \begin{pmatrix} 0 & 1 \\ 0 & -\gamma \end{pmatrix} \begin{pmatrix} x^i(t) \\ v^i(t) \end{pmatrix} + d\begin{pmatrix} 0 \\ w^i(t) \end{pmatrix}$$

Each process $z^i(t)$ is the one-dimensional Ornstein-Uhlenbeck process which at the same time is a two-dimensional Markov process with R^2 as its state space.

In the sequel we shall be dealing with the one-dimensional Ornstein-Uhlenbeck process which will be denoted

$$z(t) = \binom{x(t)}{v(t)} \quad ,$$

(dropping of the indices should not lead to any misunderstanding). Solving equation (9.5) we get by Theorem 8.2 for $t > 0$

$$z(t) = E(t)\binom{x_0}{v_0} + \int_0^t E(t-u)\, d\binom{0}{w(u)}$$

where

$$E(t) = \exp\left[t\begin{pmatrix} 0 & 1 \\ 0 & -\gamma \end{pmatrix}\right] + \sum_{n=0}^{\infty} \frac{t^n}{n!} \begin{pmatrix} 0 & 1 \\ 0 & -\gamma \end{pmatrix}^n = I + \frac{1 - e^{-\gamma t}}{\gamma} \begin{pmatrix} 0 & 1 \\ 0 & -\gamma \end{pmatrix}$$

or else

$$(9.6)\qquad z(t) = \begin{pmatrix} x_0 + \dfrac{1 - e^{-\gamma t}}{\gamma}\, v_0 + \displaystyle\int_0^t \frac{1 - e^{-\gamma(t-u)}}{\gamma}\, dw(u) \\ \\ e^{-\gamma t}\, v_0 + \displaystyle\int_0^t e^{-\gamma(t-u)}\, dw(u) \end{pmatrix}$$

Now, we easily check that

$$z(t) = \begin{pmatrix} x(t) \\ v(t) \end{pmatrix} = \begin{pmatrix} x_0 + \int_0^t v(s)\, ds \\ v(t) \end{pmatrix}$$

We introduce the following notation

$$x_1 = x_0 + \frac{1 - e^{-\gamma t}}{\gamma} v_0 , \quad v_1 = e^{-\gamma t} v_0 .$$

$$\xi = \int_0^t \frac{1 - e^{-\gamma(t-u)}}{\gamma} dw(u) , \quad \eta = \int_0^t e^{-\gamma(t-u)} dw(u) .$$

Now

$$z(t) = \begin{pmatrix} x_1 + \xi \\ v_1 + \eta \end{pmatrix}$$

Let $f(x,v)$ denotes the Gaussian density of the distribution of the random vector (ξ,η) . For given $t > 0$, x_0 and v_0

$$P\{z(t) \in A \mid z(0) = z_0\} = P\{z(t) \in A\} = \int_A f(x-x_1, v-v_1) dx\, dv ,$$

wherefrom we get that the transition probability density for the process $z(t)$ in the time t from the state $z_0 = \binom{x_0}{v_0}$ to the state $\binom{x}{v}$ is equal to

$$(9.7) \qquad f(x - x_0 - \frac{1 - \varepsilon(t)}{\gamma} v_0 , \quad v - \varepsilon(t)v_0) = p(t,z_0,z)$$

where $\varepsilon(t) = e^{-\gamma t}$. We shall show that under the proper choice of γ the density of the measure $\lambda_\beta = m \otimes \nu_\beta$ is invariant under action of (9.7). Indeed, let

$$f_2(v) = \int_{-\infty}^{\infty} f(x,v)\, dx ,$$

and

$$g_\beta(x,v) = \frac{dm \otimes \nu_\beta}{d(x,v)} = \left(\frac{\beta}{2\pi}\right)^{1/2} \exp\left(- \frac{\beta}{2} v^2\right)$$

Then

$$h(x,v) \equiv \int_{R^2} p(t,x_0,v_0;x,v) g_\beta (x_0,v_0)\, d(x_0,v_0)$$

$$= \int_R f_2(v - \varepsilon(t)v_0)(\frac{\beta}{2\pi})^{1/2} \exp(-\frac{\beta}{2} v_0^2)dv_0$$

$$= \int_R f_2(v-u)(\frac{\beta}{2\pi\varepsilon^2(t)})^{1/2} \exp(-\frac{\beta}{2\varepsilon^2(t)} u^2)du$$

On the other hard f_2 is the Gaussian density with expectation 0 and variance

(9.8) $$E(\eta^2) = \sigma_0^2 \frac{1-\varepsilon^2(t)}{2\gamma}$$

So

$$h(x,v) = g_{\beta_1}(x,v)$$

where

$$\frac{1}{\beta_1} = \frac{\varepsilon^2(t)}{\beta} + \sigma_0^2 \frac{1-\varepsilon^2(t)}{2\gamma} ,$$

Thus in order that $h = g_\beta$ it is hecessary and sufficient that

(9.9) $$\gamma D = \frac{1}{\beta} .$$

Now, let us go back to the three-dimensional model. Assuming that condition (9.9) is satisfied we get as a corollary that the equilibrium state of the ideal gas consisting of particles subject to independent Ornstein-Uhlenbeck movements will not be affected.

Finally we find the mean kinetic energy of a particle of mass 1 performing the Ornstain-Uhlenbeck motion. It is equal to

$$\frac{1}{2} E(|v(t)|^2) = \frac{3}{2} E(v^i(t))^2 = \frac{3}{2} (\varepsilon^2(t)E(v_0^2) + E(\eta^2)) ,$$

However, the observed particle in the equilibrium state has the initial velocity v_0 with the Gaussian distribution with zero mean and variance $1/\beta$. So, by (9.8)

$$\frac{1}{2} E(|v(t)|^2) = \frac{3}{2}\frac{1}{\beta} = \frac{3}{2} kT$$

what should be compared with formula (2.7).

10. Dynamics of an infinite system of particles of the ideal gas in the one-dimensional model.

In what follows we restrict our attention to the one-dimensional model. Let ξ be the equilibrium state of an infinite system of particles of the ideal gas with density ρ. By virtue of the Theorem 3.1 ξ determines the system of particles on the real line R up to a numbering. If we add to this system a particle of the same mass 1 and place it at the origin i.e. $x_0 = 0$ then all particles may be numbered so that

(10.1) $$\dots x_{-2} < x_{-1} < x_0 = 0 < x_1 < x_2 \dots$$

The important feature of the system (10.1) is that random variables $\xi_k = x_{k+1} - x_k$, $k = 0, \pm 1, \pm 2, \dots$ are independent, identically distributed with the exponential distribution with mean $1/\rho$. The system (10.1) corresponds to the initial position of particles in the equilibrium state, the additional particle being placed at the origin. Each particle x_i is subject to the uniform motion with velocity v_i, and unless the particle collides with other particles its trajectory is of the form

(10.2) $$x_i(t) = x_i + v_i t , \qquad t \geq 0 .$$

Assuming, that in our model the particles may not penetrate each other, and that the collisions are elastic we get from the principles of conservation of energy and momentum that

$$v_i + v_{i+1} = v_i' + v_{i+1}'$$

$$v_i^2 + v_{i+1}^2 = (v_i')^2 + (v_{i+1}')^2 ,$$

where v_i' denotes the velocity if i-th particle after collision.

Consequently, either $v_i = v_i'$, $v_{i+1} = v_{i+1}'$ or $v_i = v_{i+1}'$, $v_{i+1} = v_i'$. The first possibility is excluded by the lack of penetration so that in the collision the particles exchange velocities what on the figure looks as follows

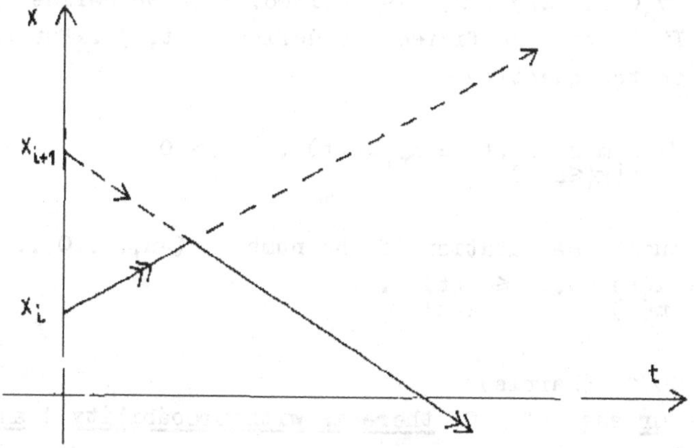

In the sequel we shall assume that the velocities are independent random variables with identical distribution ν such that

$$E(v_i) = 0 \ , \quad E(|v_i|) = \int_{-\infty}^{\infty} |x|\nu(dx) \equiv V < \infty \ ,$$

Now, we may introduce the probabilistic space of initial conditions. The vector $\zeta_k = (\xi_k, v_k)$ is a random vector on the probabilistic space (R^2, B^2, μ) where $\mu = \lambda \otimes \nu$ and

$$\lambda(dx) = I_{\langle 0,\infty)}(x) \, \rho e^{-\rho x} \, dx \ .$$

The space of initial conditions is

$$(\Omega, \mathcal{F}, P) = \prod_{-\infty}^{\infty} (R^2, B^2, \mu) \ .$$

Now, we would like to define the real trajectory $y_i(t,\omega)$ of the i-th particle. For each

$$\omega = (\ldots, \zeta_{-2}, \zeta_{-1}, \zeta_0, \zeta_1, \zeta_2, \ldots)$$

$\omega_k = \zeta_k$, $T\omega$ is defined by the formula

$$(T\omega)_k = \omega_{k+1} .$$

Suppose that $y_0(t,\omega) \equiv y(t,\omega)$ is defined. Then we define $y_k(t,\omega) = y(t,T^k\omega)$. It is then sufficient to define $y(t,\omega)$. For this purpose we introduce the quantity

$$\operatorname*{med}_{|k|\leqslant n} x_k(t) = x_{\pi(0)}(t) , \quad t > 0$$

where π is such a permutation of the number $(-n,\ldots,0,\ldots,n)$ that $x_{\pi(-n)}(t) \leqslant \ldots \leqslant x_{\pi(0)}(t) \leqslant \ldots \leqslant x_{\pi(n)}(t)$.

THEOREM 10.1. (Harris).

Part I. For any $T > 0$ *there is with probability* 1 *an* N *such that*

$$\operatorname*{med}_{|k|\leqslant n} x_k(t) = \operatorname*{med}_{|k|\leqslant N} x_k(t) \quad \text{for} \quad 0 \leqslant t \leqslant T,\ n \geqslant N .$$

Part II. If we define

$$y(t) = \lim_{n\to\infty} \operatorname*{med}_{|k|\leqslant n} x_k(t) \quad \text{for} \quad t \geqslant 0 ,$$

then with probability 1 :

1) $y_k(t)$ *are continuous on* R , $y_k(0) = x_k(0) = x_k$, $y_{k+1}(t) \geqslant y_k(t)$, *for* $t \geqslant 0$ *and* $k = 0, \pm 1, \ldots$

2) *The set theoretic unions (over* k *) of the graphs* $y_k(.)$'s *and of* $x_k(.)$'s *are identical.*

3) *If* $i \neq j$ *then* $\{t \geqslant 0 : y_i(t) = y_j(t)\}$ *does not contain any interval.*

We omit the proof of this theorem. The equilibrium state of our ideal gas at $t = 0$ has been disturbed by placing at the origin the particle x_0 and the Poissonian random measure does not correspond to this new state and that is why we may not speak about its invariance (Doob's theorem). Nevertheless one can think about the invariance of the sequence

$$(10.3) \qquad S_t\omega\ (\dots\ \zeta_{-2}(t), \zeta_{-1}(t), \zeta_0(t), \zeta_1(t),\ \dots\ )\ ,$$

where $\zeta_k(t) = (y_{k+1}(t) - y_k(t),\ v_k)$.

THEOREM 10.2 (Spitzer). For any $t \geq 0$ the sequence (10.3) is the sequence of independent vectors over $(\Omega, \mathfrak{F}, P)$, each having the distribution $\mu = \lambda \otimes \nu$.

The above theorem implies that the mapping $S_t : \Omega \to \Omega$ leaves P invariant i.e. $P = S_t P$ where, of course, $S_t P(A) = P(S_t^{-1} A)$. It turns out that for each fixed k the processes $(v_k(t),\ t \geq 0)$ and $(y_{k+1}(t) - y_k(t),\ t \geq 0)$ are stationary (Spitzer) and ergodic (Sinai).

Now we look briefly only at the relation between the process $y(t)$ and the Wiener process $w(t)$. In order to underline that y depends on the density ρ and on the mean absolute velocity V we introduce the notation $y = Y_{\rho,V}$.

THEOREM 10.3 (Spitzer [11]). Let w denotes the one-dimensional Wiener process with variance $\sigma^2 = V/\rho$ and let

$$Y_{\rho,V}^{A}(t) = \frac{Y_{\rho,V}(A^2 t)}{A}\ , \qquad 0 \leq t \leq 1\ ,\ A > 0\ .$$

Then $Y_{\rho,V}^{A}$ converges to w weakly in $C[0,1]$ as $A \to \infty$.

The more physical interpretation may be obtained with the help of the equality $Y_{\rho,V}^{A} = Y_{A\rho,AV}$.

Concluding, let us consider the modified Spitzer's model. The modification is as follows: a particle of mass $m > 1$ is placed at the origin in the ideal gas in the equilibrium state. We remind that all particles of the ideal gas are of mass 1. The particles of the gas are subject to independent uniform motions. We assume that the particles may not penetrate each other and that the collisions are elastic. Denoting the trajectory of the particle placed time t at the origin by $y_m(t)$ one proves that after a certain renorming the process $(y_m(t),\ y_m'(t))$ tends, as $m \to \infty$ weakly to the Ornstein-Uhlenbeck process (Holey [6]). The above model may also be investigated in the multi-dimensional case. On the plane it has been done in the paper [17].

References

[1] R.L.Dobrushin, On Poissona laws for distributions of particles in space, Ukrain. Math. Z. 8(1956), 127-134.

[2] J.L.Doob, Stochastic Processes, Wiley 1953, VIII, 4,5, 358 -366.

[3] A.Einstein, Investigations on the theory of Brownian motion, Dover Publ., Inc. 1956 .

[4] T.E.Harris, Diffusion with "collisions" between particles, J.Appl. Probability 2(1965), 322-338.

[5] T.E.Harris, Random measures and motion of point processes, Zeit. für Wahrsch. 2(1971), 85-115.

[6] R.Holley, The motion of a heavy particle in an infinite one dimensional gas of hard spheres", Zeit, für Wahrsch. 17(1971), 181-219.

[7] R.Holley, The motion of a large particle, Tran. Amer, Math. Soc. 144(1969), 523-534.

[8] R.Holley, A class of interactions in an infinite particle system, Advances in Math. 5(1970), 291-309.

[9] D.W.Jepsen, Dynamics of a simple many-body system of hard rods, J. of Math. Phys. 6(1965), 405-413.

[10] E.Nelson, Dynamical Theories of Brownian Motion, Princeton Univ. Press, Princeton,New Jersey 1967.

[11] F.Spitzer, Uniform motion with elastic collisions of an infinite particle system, J.Mathematics and Mechanics, 18(1969),973-990.

[12] F.Spitzer, Random processes defined through the interaction of an infinite particle system, Lecture Notes in Math. 89(1969), 201-223.

[13] F.Spitzer, Interaction of Markov Processes, Advances in Math.

[14] C.Stone, On a Theorem of Dobrushin, Ann. Math. Statist. 39(1968), 1391-1401.

[15] K.Wołkowyskij and J.Synaj, Ergodic properties of the ideal gas with infinitely many degrees of freedom, Functional Anal. and Appl. 5(1971), 19-21 (in Russian).

[16] T.E.Harris, The theory of branching processes,

[17] H.Hennion, Sur le mouvement d'une particule lourde soummise à des callisions dans un systeme infini de particules legres, Zeitschrift für Wahrsch. 25(1972/73), 123-154.

WINTER SCHOOL
ON PROBABILITY
Karpacz 1975
Springer's LNM

472

ON LÉVY'S BROWNIAN MOTION WITH SEVERAL - DIMENSIONAL TIME

By Z. Ciesielski
Institute of Mathematics of the Polish Academy of Sciences

1. Introduction.

The purpose of this lecture is to present an approach to Lévy's Brownian Motion based on Chentsov's construction [3] and on the author's note [2].

The L^2 theory and the relation of LBM (Lévy's Brownian Motion) to white noise is being discussed. The results obtained give in particular Molchan's [11] formula for the scalar product in the corresponding RKHS (Reproducing Kernel Hilbert Space). Consequently, applying the general theory of Markov fields we obtain the Markov property of LBM in the odd dimensional case.

The covariance operator for LBM is discussed in detail.

Apart from the Markov Property the approach below does not distinguish between even and odd dimension.

It is assumed further on that a probability space (Ω, F, P) rich enough is given i.e. such that there is over (Ω, F, P) an infinite sequence of independent random variables identically distributed according to $N(0,1)$ law.

The dimension d of the Euclidean space R^d is assumed to be greater than 1 always. For $a, b \in R^d$ we denote by (a,b) the usual scalar product and we put $|a| = (a,a)^{1/2}$.

The real valued Gaussian random field $\{X(a),\ a \in R^d\}$ is called LBM if it satisfies the following conditions:

$$P\{X(0) = 0\} = 1\ , \tag{1.1}$$

$$E(X(a)) = 0 \qquad \text{for}\quad a \in R^d\ , \tag{1.2}$$

$$E(X(a) - X(b))^2 = |a - b| \quad \text{for}\quad a,b \in R^d\ . \tag{1.3}$$

It follows immediately that if the Gaussian Random field $\{X(a),\ a \in R^d\}$ satisfies condition (1.1), then (1.3) is equivalent to

$$E(X(a)\,X(b)) = \tfrac{1}{2}\,(|a| + |b| - |a-b|)\ . \tag{1.4}$$

The kernel

$$K(a,b) = \frac{|a| + |b| - |a-b|}{2}\ , \qquad a,b \in R^d \tag{1.5}$$

is known as the Lévy-Schoenberg kernel. The positive definiteness of (1.5) follows from a result of Schoenberg [14] and it was used by Lévy [10] to establish the existence of Brownian motion. Another proof of this property is due to Gangolli [5].

In the Chentsov's [3] construction of LBM the structure of the kernel K plays a crucial role. That was also exploited essentially in [2]. Therefore the discussion of (1.5) is going to be our starting point.

2. The Lévy - Schoenberg kernel.

Let us start with the set of all oriented $(d-1)$ - dimensional hyper-planes in R^d. This set can be identified with the product $Z = S^{d-1} \times R$ where S^{d-1} is the $(d-1)$ - dimensional unit sphere in R^d and $R = R^1$. For given $z = (\xi;p) \in Z$ let H_z denote the corresponding hyper-plane. Clearly, if $a \in H_z$, then $(a,\xi) = p$, where (a,ξ) is the scalar product in R^d.

Now, let σ be the Borel measure on S^{d-1} invariant under the action of the orthogonal group in R^d and normalized as follows

$$\int_{S^{d-1}} |(\xi,\zeta)|\ \sigma(d\xi) = 1\ , \qquad \zeta \in S^{d-1}\ . \tag{2.1}$$

Let m be the Lebesgue measure in R and let μ be the completed product measure $\sigma \otimes m$ in Z. For given $A, A \subset R^d$, we introduce

(2.2) $$A' = \{z \in Z : H_z^* \cap A \neq \emptyset\} ,$$

where $H_z^* = \{a \in R^d : (\xi, a) = p\}$, $z = (\xi; p)$.

It can be shown that for Borel set A the set A' is analytic and therefore $\mu(A')$ is well defined. Moreover, it is well known, and not hard to see that

(2.3) $$\mu((gA)') = \mu(A')$$

for any isometry g of R^d. Consequently, if $\langle a,b \rangle = \{c \in R^d : c = t(b-a)+a,\ 0 \leq t \leq 1\}$, where $a,b \in R^d$, then $\mu(\langle 0,a \rangle') = c|a|$ with some constant c. The normalization (2.1) implies that $c = 1$ and therefore

(2.4) $$\mu(\langle 0,a \rangle') = |a| , \qquad a \in R^d .$$

It is a consequence of the Pasch's axiom in geometry that

(2.5) $$\langle a,b \rangle' \doteq \langle 0,a \rangle' \div \langle 0,b \rangle' ,$$

where $\doteq$ means equality modulo a set of μ measure zero, and $\div$ is used for the symmetric difference of two sets.

Now, denoting by I_A the indicator (characteristic function) of A we get

(2.6) $$I_{A \cap B} = \frac{1}{2} (I_A + I_B - I_{A \div B}) .$$

In particular, if $A = \langle 0,a \rangle'$ and $B = \langle 0,b \rangle'$, then integration of (2.6) with respect to μ gives

$$\mu(A \cap B) = \frac{1}{2} [\mu(A) + \mu(B) - \mu(A \div B)] ,$$

whence by (2.5) we obtain

(2.7) $$K(a,b) = \mu(\langle 0,a\rangle' \cap \langle 0,b\rangle') = \int_Z I_{\langle 0,a\rangle'} \, I_{\langle 0,b\rangle'} \, d\mu \ .$$

In our discussion still different representation of $K(a,b)$ will be needed.

Let us denote by $L^2(Z,\mu;\ C)$ and $L^2(R^d;\ C)$ the real Hilbert spaces of complex valued square integrable functions with the following scalar products respectively

$$(f,g)\mu = \int_Z \operatorname{Re}(f\bar{g})\ d\mu \ ,$$

(2.8)

$$(F,G)_d = \int_{R^d} \operatorname{Re}(F(a)\ \bar{G}(a))\ da \ .$$

The spaces $L^2(Z,\mu)$ and $L^2(R^d)$ of real valued functions are natural Hilbert subspaces of $L^2(Z,\mu;\ C)$ and $L^2(R^d;\ C)$ with the scalar products (2.8) correspondingly.

It is convenient to introduce in $L^2(Z,\mu;\ C)$ the involution

$$(If)(\xi;\ p) = f(-\xi;\ -p) \ .$$

Clearly, I leaves invariant the subspace $L^2(Z,\mu)$. Moreover, let

$$L^2_+(Z,\mu;\ C) = \left\{ f \in L^2(Z,\mu;\ C) : I\,f = f \right\}$$

$$L_+(Z,\mu) = \left\{ f \in L^2(Z,\mu) : I\,f = f \right\} \ ,$$

and let $U : L^2_+(Z,\mu) \to L^2_+(Z,\mu;\ C)$ be defined in L^2 sense as follows

$$Uf(\xi;\ p) = i \ \operatorname{sgn} p \int_{-\infty}^{\infty} e^{2\pi i p q} \operatorname{sgn} q \ f(\xi;\ q)\ dq \ .$$

The operator U has the following properties: for $f,g \in L^2_+(Z,\mu)$

(2.9) $$(f,g)_\mu = (Uf,\ Ug)\ \mu$$

$$I(Uf) = Uf, \quad \overline{Uf(\xi;\ p)} = Uf(\xi;\ -p)$$

Next is the operator $V : L_+^2(Z,\mu;\ C) \to L^2(R^d;\ C)$ defined as follows

$$Vf(a) = (c_d)^{\frac{1}{2}} \frac{F(a)}{|a|^{\frac{d-1}{2}}}$$

where $F(a) = f(\xi;\ p)$ for $a = \xi p$ and

$$c_d = \frac{\Gamma\left(\frac{d+1}{2}\right)}{\pi^{\frac{1}{2}(d-1)}} .$$

Since for $g \in L_+^2(Z,\mu;\ C)$ and $G(a) = g(\xi;\ p)$, $a = \xi p$, we have

$$\int_Z g d\mu = c_d \int_{R^d} \frac{\sigma(a)}{|a|^{d-1}}\, da ,$$

it follows that for $g, f \in L_+^2(Z,\mu;\ C)$

(2.10) $$(f,g)\mu = (Vf,\ Vg)_d .$$

Now, the last property of (2.9) implies that for $f \in L_+^2(Z;\ \mu)$ the functions $Re(VUf)$ and $J_m(VUf)$ are even and odd respectively. Consequently, for $f, g \in L_+^2(Z,\mu)$ we obtain according to (2.8) that

$$(VUf,\ VUg)_d = \int_{R^d} VUf(b)\ \overline{VUg(b)}\ db .$$

This, (2.10) and (2.9) give for $f, g \in L_+^2(Z;\ \mu)$

(2.11) $$(f,g)_\mu = \int_{R^d} VUf(b)\ \overline{VUg(b)}\ db .$$

In particular we find that $I_{\langle 0,a\rangle} \cdot \in L_+^2(Z,\mu)$ and

$$UI_{\langle 0,a\rangle}\cdot(\xi;\, p) = \frac{1}{2\pi|p|}\,(e^{2\pi i(a,p\xi)} - 1)\ ,$$

(2.12) $$VUI_{\langle 0,a\rangle}\cdot(b) = \left(\frac{c_d}{4\pi^2}\right)^{\frac{1}{2}}\ \frac{e^{2\pi i(a,b)} - 1}{|b|^{\frac{d+1}{2}}}\ .$$

This, (2.11) and (2.7) give

(2.13) $$K(a,b) = \frac{1}{2\pi\omega_{d+1}}\int_{R^d}\frac{(e^{2\pi i(a,c)} - 1)(e^{-2\pi i(b,c)} - 1)}{|c|^{d+1}}\,dc\ ,$$

where ω_{d+1} is the surface area of the unit sphere S^d in R^{d+1}, and $\omega_{d+1}c_d = 2\pi$.

THEOREM 2.1. _For given_ $f \in L^2(R^d,C)$ _let_

(2.14) $$\int_{R^d}\frac{e^{2\pi i(a,b)} - 1}{|b|^{\frac{d+1}{2}}}\,f(b)\,db = 0$$

for $a \in R^d$. _Then,_ $f = 0$.

Proof. Let for given $\varphi \in C_0^\infty(R^d)$ such that $0 \notin \operatorname{sup\,p}\varphi$ and let

$$\hat\varphi(a) = \int_{R^d} e^{2\pi i(a,b)}\,\varphi(b)\,db\ .$$

Since $\hat\varphi \in S(R^d)$ it follows that

$$\varphi(b) = \int_{R^d} e^{-2\pi i(a,b)}\,\hat\varphi(a)\,da\ ,$$

whence by $\varphi(0) = 0$ we get

(2.15) $$\int_{R^d}\hat\varphi(a)\,da = 0\ .$$

Moreover, (2.15) gives

$$\text{(2.16)} \qquad \int_{R^d} \hat{\varphi}(-a)da \int_{R^d} \frac{e^{2\pi i(a,b)} - 1}{|b|^{\frac{d+1}{2}}} f(b)\, db = 0 \ .$$

Now, let

$$g(a,b) = \hat{\varphi}(-a) \frac{e^{2\pi i(a,b)} - 1}{|b|^{\frac{d+1}{2}}} f(b) \ .$$

We find easily that

$$|g(a,b)| \leq \text{const} \ h(a)\, g(b)$$

where $h(a) = (1 + |a|)|\hat{\varphi}(-a)|$ and

$$g(b) = |f(b)| \frac{1}{(1 + |b|)|b|^{\frac{d-1}{2}}} \ .$$

Clearly $h \in L^1$, and since $f \in L^2$ it follows that $g \in L^1$. Thus, Fubini's theorem can be applied in (2.16), and this combined with (2.15) gives

$$\int_{R^d} \hat{\hat{\varphi}}(-a) \frac{f(b)}{|b|^{\frac{d+1}{2}}} db = \int_{R^d} \varphi(b) \frac{f(b)}{|b|^{\frac{d+1}{2}}} db = 0 \ ,$$

and therefore the proof is complete.

COROLLARY 2.1. *If* $f \in L^2_+(Z,\mu)$ *and*

$$\text{(2.17)} \qquad \int_{\langle 0,a\rangle} f d\mu = 0$$

for $a \in R^d$, *then* $f = 0$.

Proof. It follows from (2.11), (2.12) and Theorem 2.1.

The third representation for $K(a,b)$ we obtain directly from (2.7) and (2.10). Notice that

$$VI_{\langle 0,a\rangle}\check{}(b) = (c_d)^{\frac{1}{2}} \frac{1}{|b|^{\frac{d-1}{2}}} I_{|b|^2<(a,b)}(b) \ .$$

Consequently,

$$K(a,b) = c_d \int_{R^d} I_{|c|^2<(a,c)}(c)\, I_{|c|^2<(b,c)}(c) \frac{dc}{|c|^{d-1}} \tag{2.18}$$

COROLLARY 2.2. *If* $f \in L^2(R^d)$ *and*

$$\int_{|b|^2<(b,a)} \frac{f(b)}{|b|^{\frac{d-1}{2}}}\, db = 0$$

for $a \in R^d$, *then* $f = 0$.

This is a consequence of the fact that $V : L^2_+(Z,\mu) \to L^2(R^d)$ is an isometric isomorphism.

3. Brownian Motion and White Noise.

For a given complete probability space (Ω,F,P) let $H = L^2(\Omega,F,P)$ be the real Hilbert space of all random variables with finite second moment and with the scalar product

$$(X,Y)_\Omega = E(X,Y) \ .$$

It is assumed that there is a sequence $\{Y_n, n = 0,1,\ldots\} \subset H$ of independent random variables distributed according to $N(0,1)$ law. Moreover, let be given a complete orthonormal (CON) set $\{f_n\} \subset L^2_+(Z,\mu)$.

THEOREM 3.1 *The series*

$$(3.1) \qquad Y(a) = \sum_{n=0}^{\infty} Y_n \int_{\langle 0,a\rangle'} f_n \, d\mu \ , \qquad a \in R^d$$

converges with probability 1 uniformly on every compact set in R^d, and $\{Y(a),\ a \in R^2\}$ is a LBM .

Proof. Parseval's identity implies

$$\sum_{n=0}^{\infty} \Big(\int_{\langle 0,a\rangle'} f_n d\mu \Big)^2 = \mu(\langle 0,a\rangle') = |a|$$

and therefore by Kolmogorov's three series theorem (3.1) converges for each $a \in R^d$ with probability 1. This, (3.1) and once more Parseval's identity give

$$E(Y(a)\,Y(b)) = \sum_{n=0}^{\infty} \int_{\langle 0,a\rangle'} f_n d\mu \int_{\langle 0,b\rangle'} f_n d\mu$$

$$= \int_Z I_{\langle 0,a\rangle'} \, I_{\langle 0,b\rangle'} \, d\mu = K(a,b) \ .$$

The partial sums

$$S_n(a) = \sum_{k=0}^{n} Y_k \int_{\langle 0,a\rangle'} f_k \, d\mu$$

define a sequence of random variables with values in the Banach space

$$C_r = C(K(0,r)), \quad K(0,r) = \Big\{ a \in R^d : |a| \leq r \Big\} .$$

For given Borel set $B \subset C(K(0,r))$ we define $P_n(B) = P\{S_n(.) \in D\}$. Now, $P_n\{f(0) = 0\} = 1$ and

$$\int_{C_r} |f(a) - f(b)|^2 P_n(df) = E|S_n(a) - S_n(b)|^2$$

$$= \sum_{k=0}^{n} \left(\int_Z (I_{\langle 0,a\rangle'} - I_{\langle 0,b\rangle'}) f_k d\mu \right)^2 \leq \sum_{k=0}^{\infty} \left(\int_Z (I_{\langle 0,a\rangle'} - I_{\langle 0,b\rangle'}) f_k d\mu \right)^2$$

$$= \int_Z (I_{\langle 0,a\rangle'} - I_{\langle 0,b\rangle'})^2 \, d\mu = \int_Z |I_{\langle 0,a\rangle'} - I_{\langle 0,b\rangle'}| \, d\mu$$

$$= \mu\,(\langle 0,a\rangle' \div \langle 0,b\rangle') = |b - a| ,$$

and therefore the family of measures $\{P_n\}$, by a theorem of Berman (c.f. [1], Theorem 5.1), is uniformly tight. Now, the S_n's are sums of independent random variables with values in C_r and therefore by Theorem 3.2 of [6] they are convergent in the maximum norm of C_r with probability 1.

COROLLARY 3.1. *The realizations of LBM given by* (3.1) *are continuous with probability* 1.

Now, for a given LBM $\{X(a),\ a \in R^d\}$ let us denote by H_X the Gaussian Hilbert subspace of H generated by $\{X(a),\ a \in R^d\}$. Clearly if there are given two equivalent LBM's $\{X(a),\ a \in R^d\}$ and $\{Y(a),\ a \in R^d\}$ i.e. such that $P\{X(a) = Y(a)\} = 1$ for each $a \in R^d$, then $H_X = H_y$.

Every LBM is stochastically continuous, and this is implied by Čebyshev's inequality and by the equality $E|X(a) - X(b)|^2 = |a-b|$. Therefore, in the case of separable LBM the choice of the set of separability is irrelevant. Therefore, to each LBM $\{X(a),\ a \in R^d\}$ there is an equivalent separable LBM $\{Y(a),\ a \in R^d\}$ and consequently such that $H_x = H_y$.

THEOREM 3.2. *Let* $\{f_n,\ n = 0,1,\dots\}$ *be given ONC set in* $L_+^2(Z,\mu)$ *and let* $\{X(a),\ a \in R^d\}$ *be separable LBM. Then, there is uniquely determined ON basis* $\{X_n,\ n = 0,1,\dots\}$ *in* H_X *such that*

$$X(a) = \sum_{n=0}^{\infty} X_n \int_{\langle 0,a\rangle'} f_n \, d\mu \tag{3.2}$$

holds for all $a \in R^d$. *Moreover, the realization of* $\{X(a),\ a \in R^d\}$ *are continuous with probability* 1, *and* (3.2) *converges to* $X(a)$ *almost uniformly in* a *with probability* 1.

Proof. Suppose that we consider the family $\{X(a) : a \in D\}$ where D is countable dense subset of R^d. There is a set $N \subset \Omega$ such that for each $\omega \in \Omega - N$ the realization $X(\cdot,\omega)$ given on D can be extrapolated by a continuous realization $\tilde{X}(\cdot,\omega)$ on R^d i.e. there is continuous extension of $X(\cdot,\omega)$ from D to R^d. This can be achieved for instance by constructing a sequence of piecewise linear continuous extrapolating and almost uniformly convergent processes. Since $\{X(a), a \in R^d\}$ is separable and $\{\tilde{X}(a), a \in R^d\}$ is continuous it can be seen $\{X(a) - \tilde{X}(a), a \in R^d\}$ is separable with respect to the closed intervals. However, $X(a,\omega) = \tilde{X}(a,\omega)$ for $a \in D$ and $\omega \in \Omega - N$, and therefore $P\{X(\cdot) = \tilde{X}(\cdot)\} = 1$.

Now, let $\{Y_n, n = 0,1,\ldots\}$ be an ON basis in H_X and let

$$Y(a) = \sum_{n=0}^{\infty} Y_n \int_{\langle 0,a\rangle'} f_n \, d\mu \quad . \tag{3.3}$$

According to Theorem 3.1 $\{Y(a), a \in R^d\}$ is a continuous LBM and clearly $H_y \subset H_x$ where H_y is the Gaussian subspace generated by $\{Y(a), a \in R^d\}$. We are going to show that $H_y = H_x$. Suppose that $Y \in H_x - H_y$. Moreover, let us define for given $f \in L^2_+(Z,\mu)$

$$W(f) = \sum_{n=0}^{\infty} Y_n (f, f_n)\mu \quad . \tag{3.4}$$

It follows immediately that $E(W(f)\, W(g)) = (f,g)\mu$.

Clearly, the white noise W defined by (3.4) establishes an isometric isomorphism between $L^2_+(Z,\mu)$ and H_x. Consequently, there is an $f \in L^2(Z,\mu)$ such that $Y = W(f)$ and this element is orthogonal to H_y i.e. $E(Y(a)\, W(f)) = 0$ for $a \in R^d$. However, (3.3) and (3.4) give $Y(a) = W(I_{\langle 0,a\rangle'})$ whence

$$E\left\{Y(a)\, W(f)\right\} = \int_{\langle 0,a\rangle'} f d\mu = 0 \qquad \text{for } a \in R^d .$$

Applying now Corollary 2.1 we find that $f = 0$ and therefore $W(f) = Y = 0$. Thus, $H_x = H_y$. From this it follows that there is a unitary operator $U : H_x \to H_x$ such that $UY(a) = X(a)$ for every $a \in R^d$. Defining $X_n = UY_n$ and applying U to both sides of (3.3) we get (3.2). Uniqueness of the basis $\{X_n, n = 0,1,\ldots\}$ may be proved as follows. Suppose that in addition to (3.2) we have another basis $\{X'_n, n = 0,1,\ldots\}$ in H_x such that

$$X(a) = \sum_{n=0}^{\infty} X'_n \int_{\langle 0,a\rangle} f_n d\mu \quad , \qquad a \in R^d .$$

There is of course unitary $V : H_x \to H_x$ such that $X'_n = VX_n$ for $n = 0,1,\ldots$. Applying V to both sides of (3.2) we find that $X(a) = VX(a)$ but since $\{X(a), a \in R^d\}$ spans H_x it follows that $V = I$ and that $X'_n = X_n$.

COROLLARY 3.2. <u>It follows from</u> (3.2) <u>that every continuous LBM</u> $\{X(a,\omega) : a \in R^d, \omega \in \Omega\}$ <u>given over</u> (Ω,F,P) <u>is measurable with respect to</u> $B^d \otimes F$ <u>where</u> B^d <u>is the Borel field in</u> R^d .

Now, let a continuous LBM $\{X(a), a \in R^d\}$ be given and let $\{f_n, n = 0,1,\ldots\}$ be an ONC set in $L^2_+(Z,\mu)$. In this set up the <u>canonical white noise</u> $W_x : L^2_+(Z,\mu) \to H_x$ is defined as follows

$$W_x(f) = \sum_{n=0}^{\infty} X_n(f, f_n)_\mu \quad , \qquad f \in L^2(Z,\mu) \quad , \tag{3.5}$$

where the basis $\{X_n, n = 0,1,\ldots\}$ in H_x is given as in Theorem 3.2. Clearly,

$$(W_x(f), W_x(g)) = (f,g)_\mu \qquad \text{for } f,g \in L^2_+(Z,\mu) \, , \tag{3.6}$$

and the span of $\{W_x(f), f \in L^2_+(Z,\mu)\}$ is identical with H_x i.e. W_x is non-singular for the LBM $\{X(a), a \in R^d\}$. It follows from (3.2) and (3.5) that

$$X(a) = W_x\,(I_{\langle 0,a\rangle}\cdot) \quad , \qquad a \in R^d . \tag{3.7}$$

This equality may be understood in two ways i.e. either it means an equality of elements in $L^2(\Omega,F,P)$ for each $a \in R^d$ or there is a null set N in Ω such that for $\omega \in \Omega - N$ $X(\cdot,\omega)$ is continuous and $X(a,\omega) = W_x(I_{\langle 0,a\rangle}\cdot)(\omega)$ for all $a \in R^d$.

For the LBM (3.7) is known as Chentsov's white noise integral representation of $\{X(a), a \in R^d\}$.

It should be mentioned that the definition of W_x does not depend on the choice of the basis $\{f_n, n = 0,1,\ldots\}$. This follows from (3.7) and from the fact that $\{I_{\langle 0,a\rangle'} : a \in R^d\}$ spans $L^2_+(Z,\mu)$ (c.f. Corollary 2.1).

Representation (3.2) may now be written in term of W_x i.e. according to (3.5) $X_n = W_x(f_n)$ and therefore

$$(3.8) \qquad X(a) = \sum_{n=0}^{\infty} W_x(f_n) \int_{\langle 0,a\rangle'} f_n d\mu \ .$$

On account of Theorem 2.1 we could repeat the foregoing discussion of this section starting with (2.13) instead of (2.7). However we restrict ourself to stating the results only.

It is assumed as before that a continuous LBM is given. There will be considered two real Hilbert spaces with the scalar product given by (2.8): $L^2(R^d)$ – the space of real valued square integrable functions and $L^2_0(R^d;\, C)$ the subspace of $L^2(R^d;\, C)$ of all F such that $\overline{F}(a) = F(-a)$. The last condition is equivalent to the following two:

$$\mathrm{Re}\, F(a) = \mathrm{Re}\, F(-a) \quad \text{and} \quad -\mathrm{Im}\, F(a) = \mathrm{Im}\, F(-a)$$

Now, let for given $G \in L^2(R^d)$

$$G_{\pm}(a) = \frac{G(a) \pm G(-a)}{2} \ .$$

Then, $F = G_+ + iG_-$ is in $L^2_0(R^d;\, C)$.

Conversely, if $F \in L^2_0(R^d;\, C)$ then $G = \mathrm{Re}\, F + \mathrm{Im}\, F$ is in $L^2(R^d)$, and this establishes isometric isomorphism

$$S : L^2_0(R^d;\, C) \to L^2(R^d) \ .$$

Now, let $\{G_n, n = 0,1,\ldots\}$ be an ONC set in $L^2(R^d)$, then $\{F_n, n = 0,1,\ldots\}$, $F_n = S^{-1} G_n$ is an ONC set in $L^2_0(R^d;\, C)$.

Like in Theorem 3.2 there is unique ON basis $\{X^0_n, n = 0,1,\ldots\}$ such that

$$(3.9) \qquad X(a) = \sum_{n=0}^{\infty} X^0_n\, T\, F_n(a)$$

where for $F \in L_0^2(R^d; C)$

$$(3.10) \qquad TF(a) = \frac{1}{(2\pi\omega_{d+1})^{1/2}} \int_{R^d} \frac{e^{2\pi i(a,b)} - 1}{|b|^{\frac{d+1}{2}}} \overline{F(b)}db \ , \quad a \in R^d.$$

Correspondingly, the canonical white noise is defined as follows

$$(3.11) \qquad W_x^0(F) = \sum_{n=0}^{\infty} X_n^0 (F, F_n)_d \ , \qquad F \in L_0^2(R^d; C) \ .$$

Again W_x^0 does not depend on the choice of $\{F_n, n = 0,1,\dots\}$, and

$$(3.12) \qquad X_n^0 = W_x^0 (F_n) \ ,$$

$$(3.13) \qquad X(a) = W_x^0 (F_a)$$

where

$$(3.14) \qquad F_a(b) = \frac{1}{(2\pi\omega_{d+1})^{1/2}} \frac{e^{2\pi i(a,b)} - 1}{|b|^{\frac{d+1}{2}}} \ ,$$

and $F_a(\cdot) \in L_0^2(R^d; C)$ for all $a \in R^d$.

Representation (3.13) is well known (see e,g, [11]) and by Theorem 2.1 it is non-singular.

It is worth to point out that (3.10) and (3.14) give

$$(3.15) \qquad TF(a) = (F_a, F)_d \ ,$$

and in particular

$$(3.16) \qquad K(a,b) = (F_a, F_b)_d \ .$$

All the considerations can be reduced to real valued functions by passing from $L_0^2(R^2; C)$ to $L^2(R^d)$ via the isomorphism S .

Thus, there is unique ON basis $\{X_n^d, n = 0,1,\dots\}$ such that

(3.17) $$X(a) = \sum_{n=0}^{\infty} X_n^d \, Q \, G_n(a)$$

where for $G \in L^2(R^d)$

(3.18) $$QG(a) = \frac{1}{(2\pi\omega_{d+1})^{1/2}} \int_{R^d} \frac{\cos 2\pi(a,b)+\sin 2\pi(a,b)-1}{|b|^{\frac{d+1}{2}}} G(b)db \ ,$$

For the canonical white noise we have

(3.19) $$W_x^d(G) = \sum_{n=0}^{\infty} X_n^d \, (G, G_n)_d \quad ,$$

(3.20) $$X_n^d = W_x^d \, (G_n) \quad ,$$

and

(3.21) $$X(a) = W_x^d(G_a)$$

with

(3.22) $$G_a(b) = \frac{1}{(2\pi\omega_{d+1})^{1/2}} \cdot \frac{\cos 2\pi(a,b)+ \sin 2\pi(a,b) - 1}{|b|^{\frac{d+1}{2}}} \quad .$$

For each $a \subset R^d$ $G_a(\cdot) \in L^2(R^d)$ and for $G \in L^2(R^d)$

(3.23) $$QG(a) = (G_a, G)_d \quad .$$

In particular (3.23) gives

(3.24) $$K(a,b) = (G_a, G_b)_d$$

what for $a = b$ gives

$$\int_{R^d} \frac{1 - \cos 2\pi(a,b)}{|b|^{d+1}} \, db = \pi\omega_{d+1} \, |a|$$

which corresponds in the one-dimensional case to the well-known formula

$$\int_0^\infty \frac{1 - \cos xy}{y^2}\, dy = \frac{\pi}{2}\, |x| \, .$$

4. Reproducing Kernel Hilbert Space (RKHS) .

We are interested in the RKHS corresponding to the kernel $K(a,b)$. This space was describrd in the odd dimensional case completely by Molchan in [11]. However, we would like to consider this space for all dimensions.

The real RKHS for the kernel K is denoted by H_k and it is spanned by the family $\{K(a,\cdot),\ a \in R^d\}$ of functions with respect to the scalar product $(\cdot,\cdot)_k$ determined by

$$(4.1) \qquad (K(a,\cdot),\ K(b,\cdot))_k = K(a,b) \, .$$

Combining (4.1) and (3.24) we get

$$(4.2) \qquad (K(a,\cdot),\ K(b,\cdot))_k = (G_a,\ G_b)_d \, .$$

Substituting $G = G_b$ into (3.23) and using (4.2) we obtain

$$(4.3) \qquad (QG_a,\ QG_b)_k = (G_a,\ G_b)_d \, .$$

However, $\{G_a(\cdot),\ a \in R^d\}$ generates $L^2(R^d)$ and therefore $Q : L^2(R^d) \to H_k$ is an isometric isomorphism between these two spaces. The same can be set about $T : L^2_0(R^d;\ C) \to H_k$ where T is given as in (3.10).

Finally

$$H_k = \left\{ F : F(a) = \int_{\langle 0,a\rangle'} f d\mu \ ,\quad f \in L^2_+(Z,\mu) \right\} .$$

For a given LBM $\{X(a),\ a \in R^d\}$ there is as well natural isometric isomorphism between H_k and H_x given as follows

(4.4) $$F(a) = E(X(a)\ X)$$

where the $F \in H_k$ and $X \in H_x$ are uniquely determined by an $f \in L^2_+(Z,\mu)$ such that

$$Y = W_x(f) \quad \text{and} \quad F(a) = \int_{\langle 0,a\rangle'} f d\mu \ .$$

Now, let for $Y \in H_k$

(4.5) $$G(a) = E(X(a)\ Y)$$

It is clear that for $F,G \in H_k$ given by (4.5) and (4.4) we have

(4.6) $$(F,G)_k = (X,Y)_\Omega$$

and if e.g. $X = W^d_x(f)$, $Y = W^d_x(g)$ with $f,g \in L^2_0(R^d;\ C)$, then (4.6) gives

(4.7) $$(F,G)_k = (f,g)_d \ .$$

LEMMA 4.1. *If the real valued* $\varphi \in C^\infty_0(R^d)$ *is such that* $\varphi(0) = 0$, *then* $\varphi \in H_k$.

Proof. The property $\varphi(0) = 0$ implies

(4.8) $$\varphi(a) = \int_{R^d} (e^{2\pi i(a,b)} - 1)\ \bar{\hat{\varphi}}(b)\ db$$

$$= \frac{1}{(2\pi\omega_{d+1})^{1/2}} \int_{R^d} \frac{e^{2\pi i(a,b)} - 1}{|b|^{\frac{d+1}{2}}}\ \bar{\psi}(b)\ db$$

where $\psi(b) = (2\pi\omega_{d+1})^{1/2}\ |b|^{\frac{d+1}{2}}\ \hat{\varphi}(b)$. Since $\bar{\psi}(b) = \psi(-b)$ and $\psi \in L^2(R^d;\ C)$ it follows that ψ is in $L^2_0(R^d;\ C)$ and therefore $\varphi = T\psi \in H_k$.

THEOREM 4.1 *Let* $\varphi,\psi \in C^\infty_0(R^d)$ *be real valued and such that*

$\varphi(0) = \psi(0) = 0$. <u>Then</u>

$$\frac{1}{2\pi\omega_{d+1}} (\varphi, \psi)_k = (D^{\frac{d+1}{2}} \varphi, \psi)_d = (D^{\frac{d+1}{4}} \varphi, D^{\frac{d+1}{4}} \psi)_d$$

<u>where</u>

$$D = -\frac{1}{4\pi^2} \sum_{i=1}^{d} \left(\frac{\partial}{\partial a_i}\right)^2 , \qquad a = (a_1, \ldots, a_n) .$$

<u>Proof</u>. According to Lemma 4.1 $\varphi, \psi \in H_k$ and by (4.8)

$$(4.9) \qquad \frac{1}{2\pi\omega_{d+1}} (\varphi,\psi)_k = \int_{R^d} |b|^{\frac{d+1}{2}} \hat{\varphi}(b) \, |b|^{\frac{d+1}{2}} \overline{\hat{\psi}(b)} \, db$$

$$= \int_{R^d} |b|^{d+1} \hat{\varphi}(b) \overline{\hat{\psi}(b)} \, db$$

However, if $D^{\frac{d+1}{2}}$ and $D^{\frac{d+1}{4}}$ are understood in the generalized sense (c.f. [ ]) then

$$|b|^{\frac{d+1}{2}} \hat{\varphi}(b) = (D^{\frac{d+1}{4}} \varphi)^{\wedge} (b)$$

and

$$|b|^{d+1} \hat{\varphi}(b) = (D^{\frac{d+1}{2}} \varphi)^{\wedge} (b) .$$

This, (4.9) and Parseval's identity give the thesis of Theorem 4.1.

5. <u>The covariance operator</u>.

We assume again that on (Ω, F, P) the continuous LBM $\{X(a) : a \in R^d\}$ is given. P.Lévy [10] proved an iterated log law which implies

(5.1) $$P\left\{X(a) = 0\ \ (|a|^{\beta})\ \text{ as } |a| \to \infty\right\} = 1$$

for every $\beta > \frac{1}{2}$.

We are going to construct a real Banach space E such that $P\{X(\cdot) \in E\} = 1$. For this let be given p and q, $1 < p < \infty$, $1 < q < \infty$ and $\frac{1}{p} + \frac{1}{q} = 1$. Now, let

(5.2) $$E^* = \left\{f : |f(a)|^q (1 + |a|^{d+\frac{1}{2}})^q \in L^1(R^d)\right\}.$$

It follows from Hölder's inequality that the conjugate Banach space to E^* is

(5.3) $$E = \left\{g : |g(a)|^q \frac{1}{(1 + |a|^{d+\frac{1}{2}})^p} \in L^1(R^d)\right\},$$

if only

$$\|f\|_* = \left(\int_{R^d} (|f(a)|(1 + |a|^{d+\frac{1}{2}}))^q\, da\right)^{\frac{1}{q}}$$

and

$$\|g\| = \left(\int_{R^d} \left(|g(a)| \frac{1}{1 + |a|^{d+\frac{1}{2}}}\right)^p da\right)^{\frac{1}{p}}$$

for $f \in E^*$, $g \in E$. For convenience let us write the Hölder's inequality

(5.4) $$|(f,g)_d| \leq \|f\|_* \|g\|.$$

It is clear that E is reflexive. According to (5.1) and (5.2) we have the required property

(5.5) $$P\left\{X(\cdot) \in E\right\} = 1.$$

Our LBM is measurable and by (5.4) and (5.5) for each $f \in E^*$ we have

$$P\left\{\int_{R^d} |X(a,\omega)\ f(a)|\,da < \infty\right\} = 1 \quad ,$$

and therefore $(X,f)_d$ is Gaussian. It can be seen easily that $E[(X,f)_d] = 0$ and $(X,f) \in H_X$. Consequently, the bilinear functional

$$(5.6) \qquad R(f,g) = E[(X,f)_d\ (X,g)_d]$$

is well defined on $E^* \times E^*$. Formula (5.6) can be written in little different way. Let P_X denote the image of P under the mapping $\omega \to X(\cdot,\omega) : \Omega \to E$. Clearly, P_X is a Borel Gaussian measure on E and

$$R(f,g) = \int_E (h,f)_d\ (h,g)_d\ P_X(dh) \quad ,$$

whence by (5.4)

$$(5.7) \qquad |R(f,g)| \leq C\ \|f\|_*\ \|g\|_*$$

with

$$C = \int_E \|h\|^2\ P_X(dh) = E(\|X\|^2) < \infty .$$

It is known that square of the norm on a given Banach spaces with gaussian measure is integrable.

Now, for fixed $f \in E^*$ according to (5.7) $R(f,\cdot) \in E^{**} = E$. Thus, there is linear continuous mapping $R : E^* \to E$ with $\|R\| \leq C$ and such that $(Rf)(g) = R(f,g)$ for $f,g \in E^*$, and it is called the covariance operator for the LBM .

LEMMA 5.1. *If* $\{X(a),\ a \in R^d\}$ *is the given continuous LBM, then* $R : E^* \to H_k$ *and*

$$(5.8) \qquad Rf(a) = E((X,f)_d\ X(a)) \ ,$$

$$(Rf,\ Rg)_k = E[(X,f)_d\ (X,g)_d]$$

holds for $f, g \in E^*$.

Proof. Let us notice that $C_0^\infty(R^d)$ is dense in E^* . Moreover, for $\varphi \in C_0^\infty(R^d)$ and $f \in E^*$ we have

$$(5.9)\qquad (Rf,\varphi)_d = (Rf)(\varphi) = E((X,f)_d\,(X\varphi)_d)$$

$$= \int_{R^d} \varphi(a)\, E((X,f)_d\, X(a))\, da\ .$$

The interchange of integration is justified by the following estimate. Since $Y = (X,f)_d \in H_x$ it follows that

$$E(|YX(a)|) \leq |a|^{1/2}\, (EY^2)^{1/2}\ .$$

Since φ was arbitrary it follows by (5.9) that

$$Rf(a) = E(YX(a)) = E((X,f)_d\, X(a))\ ,$$

and this completes the proof.

We can prove now the following factorization theorem.

THEOREM 5.1. *The operator* $R : E^* \to H_k$ *can be factorized as follows:*

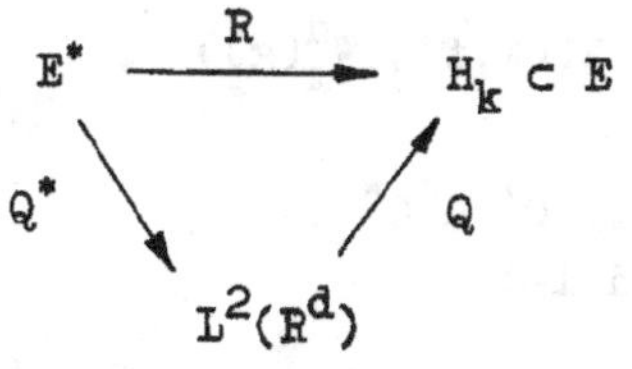

where Q *is given as in* (3.18), Q^* *is the conjugate to* Q *and the range of* Q^* *is dense in* $L^2(R^d)$.

Proof. Since $(X,f)_d \in H_x$ it follows that to each $f \in E^*$ there is a unique $h \in L^2(R^d)$ such that $(X,f)_d = W_x^d(h)$. Let this mapping be denoted Q' i.e. $Q'f = h$. According to Lemma 5.1 $R = QQ'$. Now, for $f \in E^*$ and $g \in L^2(R^d)$ we have

$$(Q'f,g)_d = E((X,f)_d \, W_x^d(g))$$

and

$$Qg(a) = E(X(a) \, W_x^d(g)) ,$$

whence for $\varphi \in C_o^\infty(R^d)$

$$(\varphi, Qg)_d = E((X,\varphi)_d \, W_x^d(g)) .$$

If $\{\varphi_n\} \subset C_o^\infty(R^d)$ is such that $\|f-\varphi_n\|_* \to 0$, then

$$P\left\{(X,\varphi_n)_d \to (X,f)_d\right\} = 1 ,$$

and since all the random variables are Gaussian it follows that

$$E((X,\varphi_n)_d - (X,f)_d)^2 \to 0 .$$

Thus,

$$(\varphi_n, Qg)_d \to (f, Qg)_d ,$$

and

$$(\varphi_n, Qg)_d \to E((X,f)_d \, W_x^d(g)) ,$$

whence $(Q'f,g) = (f,Qg)$ i.e. $Q' = Q^*$.

Now, let $g \in L^2(R^d)$ and let

$$(Q^*f,g)_d = 0 \quad \text{for} \quad f \in E^* .$$

This implies

$$(\varphi, Qg) = 0 \quad \text{for} \quad \varphi \in C_o^\infty(R^d)$$

whence $Qg = 0$, and Theorem 2.1 gives $g = 0$. Thus the range of Q^* is dense in $L^2(R^d)$.

COROLLARY 5.1. The range of R is dense in H_k .

This follows from Theorem 5.1 and the fact that Q is an isomorphism.

LEMMA 5.2. The inverse $(Q^*)^{-1}$ exists and $(Q^*)^{-1} = (Q^{-1})^*$.

Proof. Let $Q^*f_0 = 0$ where f_0 is an element from E^* . Then we have the following chain of equalities:

$$(Q^*f_0, g)_d = 0 \qquad \text{for} \quad g \in L^2(R^d) \quad ,$$

$$(f_0, Qg)_d = 0 \qquad \text{for} \quad g \in L^2(R^d) \quad ,$$

$$(f_0, QQ^*f)_d = 0 \qquad \text{for} \quad f \in E^* \quad ,$$

$$(f_0, Rf)_d = 0 \qquad \text{for} \quad f \in E^* \quad ,$$

and by Corollary 5.1.

$$(f_0, h)_d = 0 \qquad \text{for} \quad h \in H_k \quad .$$

Now, Lemma 4.1 gives

$$(f_0, \varphi) = 0 \qquad \text{for} \quad \varphi \in C_0^\infty(R^d), \quad \varphi(0) = 0 \; ,$$

and therefore $f_0 = 0$.

The domain $D((Q^{-1})^*)$ is defined as the set of all those $g \in L^2(R^d)$ for which $(g, Q^{-1}h)$ is a bounded functional in $h \in H_k$ (H_k is dense in E) .

We shall show that $\mathcal{D}((Q^{-1})^*) = \mathcal{R}(Q^*)$. If $g \in \mathcal{D}((Q^{-1})^*)$ then for some constant C

$$|(g, Q^{-1}h)_d| \leq C \, \|h\| \quad , \qquad h \in H_k \, .$$

There is unique $f \in E^*$ such that $(g, Q^{-1}h)_d = (f,h)_d$ for $h \in H_k$. Let $h = Qu$, $u \in L^2(R^d)$, then

$$(g, u)_d = (Q^*f, u)_d \qquad \text{for} \quad u \in L^2(R^d)$$

and therefore $g = Q^*f$ i.e. $g \in \mathcal{R}(Q^*)$. Now, if $g \in \mathcal{R}(Q^*)$, then for $h \in H_k$

$$(g, Q^{-1}h)_d = (Q^*f, Q^{-1}h)_d = (f, QQ^{-1}h)_d = (f,h)_d ,$$

whence $g \in \mathcal{D}((Q^{-1})^*)$.

To complete the proof let us write for $g \in \mathcal{R}(Q^*)$, $h \in H_k$

$$((Q^{-1})^*g,h)_d = (g,Q^{-1}h)_d = (g,f)_d ,$$

where $f = Q^{-1}h$, and

$$((Q^*)^{-1}g,h)_d = (Q^*(Q^*)^{-1}g,f)_d = (g,f)_d .$$

However, H_k is dense in E and therefore $(Q^{-1})^* = (Q^*)^{-1}$.

THEOREM 5.2. <u>For the given continuous LBM $\{X(a), a \in R^d\}$ the following formulas hold</u>:

(5.10) $$RF(a) = E((X,f)_d\, X(a)) , \qquad f \in E^* , a \in R^d ;$$

(5.11) $$(Rf,Rg)_k = R(f,g) , \qquad f,g \in E^* ;$$

(5.12) $$R = QQ^* ;$$

(5.13) $$(Q^*f,Q^*g)_d = R(f,g) , \qquad f,g \in E^* ;$$

(5.14) $$(f,g)_k = (R^{-1}f,g)_d , \qquad f \in \mathcal{D}(R^{-1}) , g \in H_k .$$

<u>Proof</u>. Only (5.14) requires an argument. It is a consequence of the following chain of equalities

$$(f,g)_k = (QQ^{-1}f, QQ^{-1})_k = (Q^{-1}f, Q^{-1}g)_d$$

$$= ((Q^{-1})^* Q^{-1}f, g)_d = ((Q^*)^{-1}Q^{-1}f, g)_d = ((QQ^*)^{-1}f, g)_d = (R^{-1}f, g)_d .$$

COROLLARY 5.2. *For* $\varphi \in C_0^\infty(R^d)$ *such that* $\varphi(0) = 0$ *we have*

$$R^{-1}\varphi = 2\pi\omega_{d+1} D^{\frac{d+1}{2}} \varphi . \tag{5.15}$$

This is a consequence of (5.14) and Theorem 4.1, provided that $\varphi \in \mathcal{R}(R)$. To prove this let $\psi = D^{(d+1)/2}\varphi$. It is known (c.f. [9]) that $\psi \in C^\infty(R^d)$ and

$$\psi(a) = 0\left(\frac{1}{|a|^{2d+1}}\right) \quad \text{as} \quad |a| \to \infty .$$

Thus, $\psi \in E^*$ and by (5.8)

$$R\psi(a) = E((X,\psi)_d X(a))$$

Whence we infer for $\lambda \in C_0^\infty(R^d)$, $(\lambda,1)_d = 0$,

$$(R\psi,\lambda)_d = E((X,\psi)_d(X,\lambda)) .$$

Since $\hat{\psi}(a) = |a|^{d+1}\hat{\varphi}(a)$, we get from the extended Dudley's formula for covariance (c.f. [2]) that

$$(R\psi,\lambda)_d = \frac{1}{2\pi\omega_{d+1}} \int_{R^d} \frac{(\hat{\psi}(a) - \hat{\psi}(0))(\hat{\lambda}(a) - \hat{\lambda}(0))^-}{|a|^{d+1}} da$$

$$= \frac{1}{2\pi\omega_{d+1}} \int_{R^d} \hat{\varphi}(a)\,\overline{\hat{\lambda}(a)}\, da = \frac{(\varphi,\lambda)_d}{2\pi\omega_{d+1}}$$

Thus, $\varphi = 2\pi\omega_{d+1} R\psi$ and this completes the proof.

THEOREM 5.3. *In the standard by now notation we have the following*

<u>equality</u>

(5.16) $$(X,(Q^{-1})^*g)_d = W_x^d(g) , \qquad g \in \mathcal{R}(Q^*) ,$$

<u>and</u> $\mathcal{R}(Q^*)$ <u>is dense in</u> $L^2(R^d)$.

<u>Proof</u>. For $f \in E^*$ we have $(X,f) = W_x^d(Q^*f)$ and if $g = Q^*f$, then by Lemma 5.2 $f = (Q^*)^{-1}g = (Q^{-1})^*g$. Thus ,

$$(X,(Q^{-1})^*g)_d = W_x^d(g) ,$$

whence (5.16) follows.

COROLLARY 5.3. <u>Equality</u> (5.16), <u>since</u> $\mathcal{R}(Q^*)$ <u>is dense in</u> $L^2(R^d)$, <u>can be extended in the</u> $L^2(\Omega,F,P)$ <u>sense to all</u> $g \in L^2(R^d)$. <u>Consequently</u>, <u>it makes sense to write</u>

(5.17) $$Q^{-1}X = W_x^d .$$

It is clear that for every other representation of LBM $\{X(a), a \in R^d\}$ one gets similar results.

6. <u>Markov property of LBM</u>.

We are not going to prove the Markov property but only to check sufficient conditions for this property given in [7] .

Theorem 5.2 gives for $g,f \in E^*$

$$E((X,f)_d\ (X,g)_d) = (Rf,Rg)_k = (R^{-1}Rf,Rg)_d = (f,Rg)_d$$

whence we infer for $f \in E^*$ and $h \in \mathcal{D}(R^{-1})$

(6.1) $$E((X,f)_d(X,R^{-1}h)_d) = (f,h)_d .$$

Let us consider now the LBM $\{X(a), a \in R^d\}$ a generalized process and let

$$X(\varphi) = (X,\varphi)_d \quad \text{for} \quad \varphi \in C_0^\infty(R^d) .$$

Of course, $\{X(\varphi),\ \varphi \in C_0^\infty(R^d)\}$ generates H_x .

The dual process in defined as follows

(6.2) $$\hat{X}(\varphi) = (X, R^{-1}\varphi) \qquad \varphi \in C_0^\infty(R^d) ,\quad \varphi(0) = 0 .$$

It should be clear that $\{\hat{X}(\varphi),\ \varphi \in C_0^\infty(R^d),\ \varphi(0) = 0\}$ generates H_x.

In view of (6.1) we get

(6.3) $$E(X(\varphi)\ \hat{X}(\psi)) = (\varphi,\psi)_{d+} , \qquad \psi(0) = 0 .$$

Thus, property (6.3) shows that the process $\hat{X}$ given in (6.2) is dual to X if considered on $R^d - \{0\}$. Taking into account (5.15) and (6.2) we can say that in the generalized sense

(6.4) $$\hat{X} = 2\pi\omega_{d+1}\, D^{\frac{d+1}{2}}\, X .$$

It remains to check only for $d = 2l+1$ that $(f,g)_k = 0$ for $f,g \in H_k$ with disjoint supports in $R^d - \{0\}$. By Theorem 4.1 this property is satisfied for $\varphi,\psi \in H_k$ such that $0 \notin \operatorname{supp}\varphi$ and $0 \notin \operatorname{supp}\psi$. Since the set of $\varphi = C_0^\infty(R^d)$ with $0 \notin \operatorname{supp}\varphi$ is dense in H_k we obtain the required result by suitable approximation procedure.

COROLLARY 6.1. _In the odd dimensional case the LBM_ $\{X(a), a\in R^d\}$ _is a Markov process in_ $R^d - \{0\}$ _with respect to all open sets._

References

[1] S.M.Berman, Some continuity properties of Brownian motion with time parameter in Hilbert space, Trans. Amer. Math. Soc., 3(1968), 182-198.

[2] Z.Ciesielski, Brownian Motion with a several-dimensional Time, Bull. Pol. Acad. Sci., 21(1973), 629-635.

[3] N.N.Chentsov, Mnogoparametricheskoye brounovskoye dvizhenie Lévy i obobshchenyi belyi shum, Teoria Veroyatn. i Ee Prim., 2(1957), 281-282.

[4] N.N.Chentsov, Sluchainye pola Lévy, ibid., 13(1968), 152-155.

[5] R.Gangolli, Positive definite kernels on homogeneous spaces and certain stochastic processes related to Lévy's Brownian motion of several parameters, Ann. Inst. Henri Poincaré (Section B), 3(1967), 121-225.

[6] K.Ito, M.Nisio, On the convergence of sums of independent Banach space valued random variables, Osaka J. Math. 5(1968), 35-48.

[7] G.Kallinapur and V.Mandrekar, The Markov property for generalized Gaussian random fields, Ann. Inst. Fourier 24(1974), 143-167.

[8] H.P. Mc Kean, Jr., Brownian motion with a several-dimensional time, Teoria Veryoat. i Ee Prim., 8(1963), 357-378.

[9] N.S.Landkov, Osnovy sovremionnoi teori potenciaua, Moskva 1966.

[10] P.Lévy, Processus stochastiques et monvement Brownian, Gauthier-Villars, Paris, 1947.

[11] G.M.Molchan, On some problems concerning Brownian motion in Lévy's sense, Theory of Prob. Appl. 12(1967), 747-755.

[12] - , Characterization of Gaussian fields with Markovian property, Dokl. Akad. Nauk SSSR, 197(1971), 784-787.

[13] L.D.Pitt, A Markov Property for Gaussian Processes with a Multidimensional Parameter, Arch. Rational Mech. Analysis, 43(1971), 367-391.

[14] I.J.Schoenberg, Metric spaces and positive definite functions Trans. Amer. Math. Soc. 44(1938), 811-841.

WINTER SCHOOL
ON PROBABILITY
Karpacz 1975
Springer's LNM

472

CONVERGENCE OF OBSERVABLES

By R.Jajte
Łódź University

Introduction

The evolution of a quantum - mechanical system may be described either by changes the system states (Schroedinger picture) or by changes of the observables (Heisenberg picture). In particular, to describe the asymptotic behaviour of the system it is reasonable to investigate various types of convergence of spectral or semispectral measures. Spectral measures correspond to self-adjoint operators (physical quantities, observables). Semispectral measures may be in the natural way interpreted in quantum statistics (in the theory of decision functions) as randomized strategies. More precisely, randomized strategies are described by commutative semispectral measures (see [2]). It should be noted that recently Holevo has given an example showing that noncommutative semispectral measure may serve as a (generalized, nonrandomized) strategy allowing to decrease the risk in comparison with randomized strategies ([2], p.148).
In § 1 we shall consider the convergence of observables (self-adjoint operators) in terms of semispectral measures theory. As the starting point to our discussion we shall use considerations to be found in [3]. § 2 will present the application of various types of the convergence of observables to the investigation of series of observables (generalized observables) in some tensor products. § 2 can thus be treated as the illustration of efficiency of methods given in § 1 .

I am indebted to Dr. E.Hensz for valuable comments and substantial help during the preparation of the paper.

§ 1. Convergence of observables.

1.1. Let H be a complex, separable Hilbert space. In the algebra L(H) of bounded linear operators in H the following basic types of convergence are considered : uniform, strong and weak. Let

$$(*) \qquad A_n \to A$$

denote convergence in any of the above meanings. The condition (*) makes no sense for unbounded operators. Let C denote the space of functions continuous and bounded on the real line R . The condition

$$(**) \qquad f(A_n) \to f(A) \qquad \text{for any } f \in C$$

makes sense for each self-adjoint operator (bounded or not) and for any type of convergence listed above. If the arrow "→" denotes the weak (resp. strong, uniform) convergence, then (**) will be read: the sequence A_n is C - weakly (resp. C-strongly, C-uniformly) convergent to the operator A . The spectral Theorem allows us to reduce the question of convergence of unbounded self-adjoint operators to the investigation of the corresponding spectral measures. Some theorems may be more naturally formulated in terms of semispectral measures. For this reason it seems desirable to reformulate (**) so that it could be read in terms of semispectral measures. Let us recall now definitions of the notions involved.

1.2. DEFINITION. Let B denote the Borel σ - algebra of subsets of R and $L^+(H)$ the set of bounded, symmetric and positive operators. By a semispectral measure we shall mean a mapping $F : B \to L^+(H)$ satisfying the conditions:

1^0 $F(R) = I$

2^0 for any $x \in H$, the function $m_x(\cdot) = (F(\cdot)x, x)$ is a measure on B .

In case when for any pair $Z_1, Z_2 \in B$ we have $F(Z_1)\, F(Z_2) = F(Z_2)\, F(Z_1)$, the semispectral measure is called commutative. Moreover, if $F(Z_1 \cap Z_2) = F(Z_1)\, F(Z_2)$ holds for any $Z_1, Z_2 \in B$, F is called a spectral measure. F(Z) are then orthogonal projectors in H .

The well known theorem of Naimark (see e.g. [1]) gives the relationship between spectral and semispectral measures. Namely, if F is a semispectral measure in H, then there exists a Hilbert overspace H_1 of the space H and a spectral measure E in H_1 such that $F(\cdot) = P\,E(\cdot)P$ where P is the projection of H_1 onto H.

A semispectral measure F generates a family of measures on R : $\{m_x(\cdot) = (F(\cdot)x, x),\ x \in H\}$. The following lemma is easy to prove:

1.3. LEMMA (cf. [3]). <u>For a system $\{m_x, x \in H\}$ of measures on R there exists a semispectral measure F such that $(F(\cdot)x, x) = m_x(\cdot)$ for any $x \in H$ if, and only if the family of measures $\{m_x, x \in H\}$ satisfies the conditions</u>

(α) $$m_x(R) = \|x\|^2$$

(β) $$m_{x+y} + m_{x-y} = 2(m_x + m_y)$$

(γ) $$m_{\lambda x} = |\lambda|^2\, m_x$$

<u>for every $x \in H$ and an arbitrary scalar λ. The family of measures $\{m_x, x \in X\}$ derives from a spectral measure F if and only if it satisfies the conditions (α), (β), (γ) and, additionaly, the condition:</u>

(δ) $$m_x = \sum_{k=1}^{\infty} |\beta_{x,e_k}|^2$$

<u>for any $x \in H$ and orthonormal basis (e_k), where</u>

$$\beta_{x,y} = \frac{1}{4}\,(m_{x+y} - m_{x-y} + i m_{x+iy} - i m_{x-iy})\ .$$

1.4. LEMMA <u>Let F_n be a sequence of semispectral measures such that for every $x \in H$ the sequence of measures $m_x^n(\cdot) = (F_n(\cdot)x, x)$ is weakly convergent to a measure m_x. Then the system of measures $\{m_x, x \in H\}$ derives from some semispectral measure F. Obviously the system $\{m_x, x \in H\}$ uniquely determines the semispectral measure.</u>

1.5. <u>Remark</u>. If (F_n) is a sequence of spectral measures, then passing to the limit (as in Lemma 1.3) does not in general give a spectral measure. It can be seen in the following example: for $\|x\| = 1$, let $\hat{x}$ denote the projection on the x - axis, i.e.

$\hat{x}h = (h,x)x$ for $h \in H$. Put $f_n = (e_0 + e_n) / \sqrt{2}$ where (e_n) is an orthonormal basis in H. It suffices to consider the spectral measures of the operators $\hat{f}_n$.

1.6. If F is a semispectral measure, then a one-parameter family of contractions in H defined by

$$F^{\wedge}(t) = \int e^{itu}\, F(du), \qquad t \in R \tag{1}$$

is called the Fourier transformof the semispectral measure F. The transform $F^{\wedge}$ uniquely determines the semispectral measure. When F is a spectral measure, $(F^{\wedge}(t),\ t \in R)$ is obviously a group of unitary operators.

1.7. Going back to the condition (**) and using the Spectral Theorem, we can rewrite (**) in the form

$$\int f(u)\, E_n(du) \to \int f(u)\, E(du) \qquad \text{for every } f \in C \tag{$_*{}^*{}_*$}$$

in the suitable topology.
The relation $({}_*{}^*{}_*)$ suggests the following definition of convergence of a sequence of semispectral measures.

1.8. DEFINITION. It is said that a sequence (F_n) of semispectral measures is C - weakly (resp. C - strongly, C - uniformly) convergent to a semispectral measure F, if for any function $f \in C$, the sequence of integrals $\int f(u)\, F_n(du)$ is weakly (resp. strongly, uniformly) convergent to the integral $\int f(u)\, F(du)$.

1.9. PROPOSITION. (α) The topology of C - weak convergence of semispectral measures is metrizable. The space of all semispectral measures with the metric

$$d(F,G) = \sum_{k=1}^{\infty} \frac{1}{2^k}\, \rho(F_{x_k}, G_{x_k}),$$

where $F_g(.) = (F(.)g, g)$, ρ is the Levy metric and (x_k) is a dense sequence in the unit sphere of H, is complete.

(β) For C - weak convergence of a sequence (F_n) of semispectral measures, the weak convergence of the sequences of measures $(F_n(.)x, x) \xrightarrow{\text{weakly}} m_x$ for x's from a dense subset of H is sufficient.

Proof. Obviously the C - weak convergence of semispectral measure implies the convergence in metric d. Conversely let us first notice the inequality

$$(2) \qquad \rho(F_h, F_g) \leq 2\|h - g\| .$$

In fact, we have for every $\eta > 0$ and $x \in R$

$$(3) \quad F_h((-\infty, x - \eta)) - \eta \leq F_h((-\infty, x)) \leq F_h((-\infty, x + h)) + \eta$$

Put $\delta = \eta + 2\|g - h\|$. Then (3) implies the inequalities:

$$F_h((-\infty, x - \delta)) - \delta \leq -2\|g - h\| + F_h((-\infty, x))$$

and

$$F_h((-\infty, x)) + 2\|g - h\| \leq F_h((-\infty, x + \delta)) + \delta$$

Since for every $Z \in B$

$$-2\|g - h\| + F_h(Z) \leq$$

$$F_g(Z) - (F(Z)(g - h), g) + F_h(Z) + (F(Z)h, g - h) \leq 2\|g - h\| + F_h(Z) ,$$

we obtain for each x

$$F_h((-\infty, x - \delta)) - \delta \leq F_g((-\infty, x)) \leq F_h(-\infty, x + +\delta)) + \delta .$$

Hence

$$\rho(F_h, F_g) \leq \delta = \eta + 2\|g - h\|$$

and thus $\rho(F_h, F_g) \leq 2\|g - h\|$.

Part (β) of the proposition follows easily from the inequality (2). Then part (α) follows from Lemma 1.4.

1.10. THEOREM. The sequence (F_n) of semispectral measures is C - weakly convergent to the semispectral measure F if and only if the sequence of the Fourier transforms $F_n\hat{}(t)$ is, for any $t \in R$, weakly convergent to the operator-valued function $F(t)$ such that for any $x \in H$, $(F(t)x, x)$ is a characteristic function of a measure

m_x on the real line.

The proof follows immediately from the definition of the C - weak convergence and Lemma 1.4.

1.11. Remark. If (F_n) are spectral measures, then the C - weak limit need not be a spectral measure (cf. example 1.5).

1.12. THEOREM. The sequence (F_n) of semispectral measures is C - strongly convergent to the semispectral measure F if and only if the sequence of the Fourier transforms $F_n^\wedge(t) = \int e^{itu} F_n(du)$ is strongly convergent to the Fourier transform $\varphi(t) = \int e^{itu} F(du)$ for any $t \in R$.

Proof. It is enough to show sufficiency. From the assumption the sequence of the measures $m_x^n(.) = (F_n(.)x, x)$ is weakly convergent for any $x \in H$, and therefore tight. Thus there exists $c > 0$ such that $(F_n(|u| \geq c)x, x) < d$ for $n = 1,2,\ldots$. Let $f \in C$ and $W(u) = \sum_k a_k e^{it_k u}$ be a trigonometric polynomial such that $\sup_{|u|\leq c} |f(u) - W(u)| \leq d$, $\|W\| \leq 2\|f\|$. Then we have

$$\int_R f(u) F_n(du)x =$$

$$\int_R W(u) F_n(du)x + \int_{|x|\leq c} [f(u) - W(u)] F_n(du)x + \int_{|x|\geq c} [f(u) - W(u)] F_n =$$

$$a_n + b_n + c_n \quad ,$$

where

$$a_n = \int W(u) F_n(du)x = \sum_k a_k F_n^\wedge(t_k)x \to \sum_k a_k F^\wedge(t_k)x \quad .$$

A standart estimation of the integrals b_n and c_n gives

$$\|c_n\| \leq 3\|f\| \sup_{\|y\|=1} (F(|u| \geq c)x, x)^{\frac{1}{2}} (F(|u| > c)y, y)^{\frac{1}{2}} <$$

$$3\|f\| \sqrt{d} \quad \|b_n\| \leq d\|x\| \quad ,$$

what ends the proof.

1.13. Remark. Theorem 1.12. remains still valid if we substitute everywhere spectral for semispectral measures (cf. [3]).

§ 2. Convergence of Sums in Tensor Product.

2.1. Let $A_1,\dots,A_n$ be self-adjoint operators and $E_1,\dots,E_n$ the corresponding spectral measures i.e. $A_k = \int u\, E_k(du)$ for $k = 1,2,\dots,n$. Consider the tensor product V^t of the Fourier transforms (unitary groups) e^{itA_k} :

$$V^t = e^{itA_1} \otimes \dots \otimes e^{itA_n} = \int e^{itu}\, E_1(du) \otimes \dots \otimes \int e^{itu}\, E_n(du)$$

Clearly, $(V^t, t \in R)$ is a unitary group in the n-th tensor power $H^{\otimes n}$. Denote by $A = \int u\, E(du)$ the self-adjoint operator being an infinitesimal generator of the group $(V^t, t \in R)$. This operator will be called a tensor sum of operators $A_1,\dots,A_n$ and denoted $A = \sum_{k=1}^{n}{}^{\otimes} A_k$. The corresponding spectral measure E of the operator A will be also called, for the sake of symmetry, a tensor sum of spectral measures $E_1,\dots,E_n$ and denoted $E = \sum_{k=1}^{n}{}^{\otimes} E_n$.

2.2. We shall now generalize the notion of a tensor sum of observables onto semispectral measures (generalized observables) . Let $F_1,\dots,F_n$ be semispectral measures in H . By Naimark theorem there exist overspaces H_j and spectral measures E_j in H_j , $j = 1,2,\dots,n$, such that $F_j = P_j E_j P_j$, where P_j is projection of H_j onto H . Let U_j^t , V_j^t $(j = 1,2,\dots,n,)$ be the Fourier transforms of spectral measures E_j and semispectral measures F_j , respectively. Denote by E the spectral measure corresponding to the unitary group $U_1^t \otimes \dots \otimes U_n^t$. The formula

$$F(.) = (P_1 \otimes P_2 \otimes \dots \otimes P_n)\, E(.) \tag{4}$$

defines a semispectral measure in $H^{\otimes n}$. $F(.)$ is independent of the overspaces H_j and the spectral measures E_j and is uniquely determined by the semispectral measures $F_1,\dots,F_n$. Indeed, from the formula

$$(P_1 \otimes \dots \otimes P_n)(U_1^t \otimes \dots \otimes U_n^t) = V_1^t \otimes \dots \otimes V_n^t \tag{5}$$

it follows that $V_1^t \otimes \dots \otimes V_n^t$ is the Fourier transform of the semispectral measure $F(.)$. The semispectral measure F will be called a tensor sum of the semispectral measures $F_1,\dots,F_n$ and denoted

$F = \sum_{k=1}^{n}{}^{\otimes} \; F_k$.

Thus in particular, we have

$$(6) \qquad \left(\int e^{itu} \sum_{k=1}^{n}{}^{\otimes} F_k(du) \bigotimes_1^n x_k, \; \bigotimes_1^n x_k \right) =$$

$$\left(\bigotimes_1^n \int e^{itu} F_k(du) \bigotimes_1^n x_k, \; \bigotimes_1^n x_k \right) = \prod_{k=1}^{n} \int e^{itu} \left(F_k(du)\, x_k, \, x_k \right) .$$

2.3. Let us consider a sequence of copies of H and a unit vector e in H .
Take a space $E \subset \prod_1^\infty H$ consisting of sequences $x = (x_1, x_2, \ldots)$ such that $x_j = e$ for all but finite number of j's . Define a positive definite kernel K over E putting $K(x,x') = \prod_i (x_i, x_i')$.
Let C^E denote the set of all complex functions defined on E and vanishing everywhere but on finite set of arguments. We shall introduce in C^E an hermitian form

$$\langle f, f' \rangle = \sum_{x,x' \in F} K(x, x')\, f(x)\, \overline{f'(x')} \quad .$$

Completing the quotient space

$$C^E / \{ \langle f, f \rangle = 0 \}$$

with respect to the norm $\langle f, f \rangle^{\frac{1}{2}}$ we obtain a Hilbert space (tensor product) which we shall denote by $\bigotimes_1^{\infty}{}^{e} H$.

2.4. Consider a pair $(Y,y) = (\bigotimes_1^\infty{}^e H \, , \bigotimes_1^\infty e)$ where $y = \bigotimes_1^\infty e$ denotes the canonical image in Y of the sequence $(e,e,\ldots)$. The pair $(\bigotimes_1^\infty{}^e H \, , \bigotimes_1^\infty e)$ may be treated as an analogue of a Cartesian product of probability spaces $(\prod_1^\infty \Omega \, , \prod_1^\infty p)$. It is worth to note that, roughly speaking, in the space $\bigotimes_1^\infty{}^e H$ there exist no states being tensor counterparts of the Cartesian measure product, except the pure state $\bigotimes_1^\infty e$.

2.5. In the sequel we shall consider series ob observables (generalized observables) of the form

$$\sum_{k=1}^{\otimes} F_k \tag{7}$$

where F_k are semispectral (spectral) measures. The sum of the series (7) will be understood as a limit (in a suitable topology) of the sums

$$S_N(.) = \Big(\sum_{k=1}^{N}{}^{\otimes} F_k(.)\Big) +^{\otimes} D^N(.) \ , \tag{8}$$

where D^N denotes the spectral measure of the zero-operator in the space $Y^N = \bigotimes_{N+1}^{\infty}{}^{e} H$.

2.6. In the sequel we shall use one more concept of convergence of semispectral measures. It is said that the sequence (F_n) is convergent in law at the state x_0 if the sequence of measures on R , $(F_n(.)x_0, x_0)$ is weakly convergent to some probability measure on R .

Consequently, the C - weak convergence of (F_n) is the convergence in law at every state x .

We shall prove now the analogue of the classic Levy's theorem for series of independent random variables.

2.7. THEOREM. <u>For the series</u> $\sum_{k=1}^{\infty}{}^{\otimes} F_k$ <u>of semispectral measures</u>

(i_1) <u>the</u> C - <u>weak convergence in</u> $\bigotimes_1^{\infty}{}^{e} H$ <u>and</u>

(i_2) <u>the convergence in law at the state</u> $y = \bigotimes_1^{\infty} e$ <u>are equivalent.</u>

<u>Proof</u>. The implication $(i_1) \to (i_2)$ is obvious. Let now the series (7) be convergent in law at the state $y = \bigotimes_1^{\infty} e$. The n-th sum (8) of the series (7) will be shortly denoted by S_N . Consider the sequence of independent random variables ξ_k with distributions $(F_k(.)e, e)$, $k = 1,2,\ldots$. The quadratic form $(\hat{S_N} y, y)$ equals

$$(\hat{S_N} y, y) = \int e^{itu} (S_N(du)y, y) = \tag{9}$$

$$\prod_{k=1}^{N} \int e^{itu} (F_k(du)e, e) = \prod_{k=1}^{N} E(e^{it\xi_k}) .$$

Thus the convergence of the series $\sum_{k=1}^{\infty} \xi_k$ is equivalent to the convergence of the quadratic forms $(S_N y, y)$ as $N \to \infty$, i.e. to the convergence in law of the series $\sum_{k=1}^{\infty \otimes} F_k$ at the state y. Therefore, in particular, the infinite product

$$\prod_{k=m}^{\infty} \int e^{itu} (F_k(du)e, e)$$

converges to the Fourier transform of some measure on R for every m

Let now $(\varphi_0, \varphi_1, \varphi_2, \dots)$ be an orthonormal basis in H and $\varphi_0 = e$. Then the elements $\overset{\infty}{\underset{1}{\otimes}} x_k$, where $x_k = e$ for almost all k's, and for other k's x_k are equal to the elements of the basis $(\varphi_1, \varphi_2, \dots)$, form the orthogonal basis in $Y = \overset{\infty}{\underset{1}{\otimes}}{}^e H$.

Let $\overset{\infty}{\underset{1}{\otimes}} x_k$, $\overset{\infty}{\underset{1}{\otimes}} x_k'$ be two elements of our basis in Y. Therefore there exists k_0 such that for $k \geq k_0$ we have $x_k = x_k' = e$. Therefrom it follows that as $N \to \infty$, the integrals

$$\int e^{itu} (S_N(du) \overset{\infty}{\underset{1}{\otimes}} x_k , \overset{\infty}{\underset{1}{\otimes}} x_k')$$

are convergent to the infinite product

$$\prod_{k=1}^{k_0} \int e^{itu} (F_k(du) x_k, x_k') \prod_{k_0+1}^{\infty} \int e^{itu} (F_k(du) e, e) .$$

Hence it follows that the measures on R $(S_N(.)x, x)$, $N = 1,2,\dots$, are weakly convergent to some measures m_x for vectors x from a set dense in Y (a linear combination of vectors of the basis). An application of the proposition 1.9. completes the proof.

Thus the convergence in law of S_N at only one state y implies the convergence in law of S_N at every state (to a semispectral measure).

When the components F_k of the series (7) are semispectral measures (orthogonal), the sum of the series need not be a spectral measure. It is, however, worth noticing that if $\sum_1^{\otimes} F_k$ is a spectral measure, then the C - weak and C - strong convergence of the series are equivalent (cf. [3]).

2.8. The convergence in law of the series $\sum_1^{\otimes} F_k$ may be shown using Kolmogorov Three Series Theorem. We shall formulate it here in terms of self-adjoint operators i.e. we shall speak of the convergence of the series $\sum_1^{\otimes} A_k$.

The n-th sum of the series is understood as

$$S_N = \sum_1^{N}{}^{\otimes} A_k +^{\otimes} \theta_N , \tag{10}$$

where θ_N is a zero-operator in $\bigotimes_{N+1}^{\infty}{}^{e} H$. For a self-adjoint operator A and a state x put

$$\mu_x(A) = (Ax, x) , \qquad v_x A = (Ax, Ax) - (Ax, x)^2$$

(if they exist).

2.9. PROPOSITION. (α). <u>If</u> $\sum_k v_e(A_k)$, $\sum_k \mu_e(A_k)$ <u>are convergent, then the series</u> $\sum_k^{\otimes} A_k$ <u>is convergent in law at the state</u> $y = \bigotimes_1^{\infty} e$

(β) <u>The series</u> $\sum_k^{\infty}{}^{\otimes} A_k$ <u>is convergent in law at the state</u> y <u>if and only if the series;</u>

(i) $$\sum_k \|F_k(|u| \geq 1)e\|^2 ,$$

(ii) $$\sum_k v_e(\bar{A}_k) ,$$

(iii) $$\sum_k \mu_e(\bar{A}_k) ,$$

<u>where</u> $\bar{A}_k = \int_{|u|\leq 1} u\, F_k(du)$, <u>are convergent.</u>

<u>Proof</u>. It suffices to put $p_k(\cdot) = \|F_k(\cdot)\, e\|^2$ and consider the sequence of independent random variables $\xi_1, \xi_2, \ldots$ with distribu-

tions $p_1, p_2, \ldots$, respectively. Then the proposition follows from the well known classic theorems on series of independent random variables.

References

[1] N.Dunford and J.T.Schwartz, Linear Operators, Part II, New York, London 1963.

[2] A.S.Holevo, An analoque of the theory of statistical desicions and the noncommutative probability theory, Trudy Moskov. Mat. Obsc: 26 (1972), 133-149.

[3] R.Jajte and B.Nowak, On convergence of observables, To appea in Rep. Mathematical Phys.

[4] M.Loève, Probability Theory, New York, London, 1960.

[5] G.W.Mackey, The Mathematical Foundations of Quantum Mechanics, New York, Amsterdam 1963.

[6] E.Prugovecki, Quantum Mechanics in Hilbert Space, New York, London 1971.

WINTER SCHOOL
ON PROBABILITY
Karpacz 1975
Springer's LNM

472

A LIMIT THEOREM FOR TRUNCATED RANDOM VARIABLES

By Z. Jurek
Wrocław University

Let H be a separable real Hilbert space. For every positive r we define a mapping T_r from H onto itself by means of the formula $T_r x = 0$ whenever $\|x\| \le r$ and $T_r x = (1 - \frac{1}{\|x\|})x$ in the remaining case. The problem we study is enunciated as follows: suppose that $\{X_n\}$ is a sequence of independent identically distributed H - valued random variables and assume that $\{r_n\}$ is an increasing sequence of positive numbers and $\{a_n\}$ is a sequence of vectors from H such that the distribution of

$$(1) \qquad \sum_{k=1}^{n} T_{r_n} X_k + a_n$$

converges (weakly) to a measure μ ; what can be said about the limit measure μ ? Converting this to a problem involving only probability measures we ask which measures μ can arise as limits of sequences

$$\lambda_n^{*n} * \delta_{a_n} \qquad (n = 1,2,\ldots)$$

where the power is taken in the sense of convolution, δ_a denotes the probability measure concentrated at the point a and λ_n is given by the formula

(2) $$\lambda_n(E) = \lambda(T_{r_n}^{-1}(E)) \qquad (n \quad 1,2,\ldots),$$

λ being an arbitrary probability measure on H. It is clear that for non-degenerate random variables $X_1, X_2, \ldots$ the existence of the limit distribution of (1) yields the relation $r_n \to \infty$ or, in other words that the random variables $T_{r_n} X_k$ $(k = 1,2,\ldots,n\,;\; n = 1,2,\ldots)$ form a uniformly infinitesimal triangular array. Hence it follows that the limit distribution μ is infinitely divisible. Consequently, its characteristic function $\hat{\mu}$ is of the form

(3) $$\hat{\mu}(y) = \exp\left\{i(a,y) - \frac{1}{2}(Qy,y) + \int_{H-\{0\}} \left(e^{i(x,y)} - 1 - \frac{i(x,y)}{1+\|x\|^2}\right) M(dx)\right\},$$

where a is a fixed element of H, Q is an S - operator and M is a σ - finite measure with finite mass outside every neighborhood of the origin and

$$\int_{\|x\|\leq 1} \|x\|^2 M(dx) < \infty .$$

Moreover, this representation is unique ([1], Theorem 4.10, Chap. VI).

Our first aim is to determine the parameters Q and M corresponding to non-degenerate limit distributions μ. We have seen that $r_n \to \infty$ is this case. Moreover, introducing the notation

(4) $$b_n = \int_{\|x\|\leq 1} x\, \lambda_n(dx),$$

(5) $$\Theta_n = \lambda_n * \delta_{-b_n}$$

and

(6) $$\nu_n = e(\Theta_n)^{*n} * \delta_{nb_n},$$

where for any finite measure m on H the compound Poisson distribution $e(m)$ is given by the formula

$$e(m) = e^{-m(H)} \sum_{k=0}^{\infty} \frac{m^{*k}}{k!} ,$$

we infer that

(7) $$\mu = \lim_{n\to\infty} \nu_n * \delta_{a_n}$$

([1], Theorem 6.2 and Corollary 6.1). Of course, ν_n is infinitely divisible and has no Gaussian component. Moreover, the characteristic function $\hat{\nu}_n$ is of the form

$$\hat{\nu}_n(y) = \exp\left\{i(c_n,y) + \int_{H-\{0\}} (e^{i(x,y)} - 1 - \frac{i(x,y)}{1+\|x\|^2}) M_n(dx)\right\} ,$$

where $M_n = n\,\theta_n$ and θ_n is determined by (5). Further. taking into account (2) , we have the formula

(8) $$M_n(E) = n\lambda\,(T_{r_n}^{-1}(E + b_n)) .$$

By (7) and the convergence theorem ([1], Theorem 5.5 , Chap. VI) we have

(9) $$\lim_{n\to\infty} M_n = M$$

outside every closed neighborhood of the origin and

(10) $$\lim_{\varepsilon\to 0} \overline{\lim_{n\to\infty}} \int_{\|x\|\le\varepsilon} (x,y)^2 M_n(dx) = (Qy,y) .$$

Given a Borel subset $\mathfrak{U}$ of the unit sphere of H and an interval I of the positive half-line $(0,\infty)$, we denote by $[\mathfrak{U},I]$ the subset $\left\{x\colon \frac{x}{\|x\|} \in \mathfrak{U} , \|x\| \in I\right\}$ of H . It is evident that

(11) $$T_r^{-1}[\mathfrak{U}, I] = [\mathfrak{U}, I+r] .$$

We have assumed that the limit measure μ is non-degenerate. Consequently, $r_n \to \infty$ and $\lambda_n \to \delta_0$. Thus, by (4),

(12) $$b_n \to 0 \ .$$

LEMMA 1. <u>For every limit measure</u> μ <u>the</u> S - <u>operator</u> Q <u>vanishes identically</u>. <u>In other words</u>, μ <u>has no Gaussian component</u>.

<u>Proof</u>. Obviously, it suffices to consider non-degenerate measures μ . Since, by (8)

$$M_n(\{x: \|x\| \leq \varepsilon\}) \leq n\lambda(T_{r_n}^{-1}\{x: \|x\| \leq \varepsilon + \|b_n\|\})$$

$$\leq M_n(\{x: r_n - \|b_n\| \leq \|x\| \leq \varepsilon + \|b_n\| + r_n\}) \quad ,$$

we infer, by virtue of (9) and (12) that for every positive ε

$$M_n(\{x: \|x\| \leq \varepsilon\}) \to 0 \ .$$

Consequently, by virtue of (10),

$$(Qy,y) \leq \lim_{\varepsilon\to 0} \lim_{n\to\infty} \varepsilon^2 \|y\|^2 M_n(\{x: \|x\| \leq \varepsilon\}) = 0 \quad ,$$

which completes the proof of the Lemma.

LEMMA 2. <u>If the limit distribution</u> μ <u>is non-degenerate, then</u>

$$r_{n+1} - r_n \to 0 \ .$$

<u>Proof</u>. Suppose the contrary and denote by q a limit point of the sequence $\{r_{n+1} - r_n\}$ with $0 < q \leq \infty$. Consider the set $[\mathfrak{U},I]$ for which the boundaries $\partial[\mathfrak{U}, I+kq]$ $(k = 0,1,\ldots)$ have the M - measure zero. Here $[\mathfrak{U}, I+\infty]$ denotes the empty set. By (8) and (11),

$$M_{n+1}([\mathfrak{U}, I] - b_{n+1}) = (n+1)\ \lambda([\mathfrak{U}, I+r_{n+1}])$$

$$= \frac{n+1}{n} M_n\ ([\mathfrak{U}, I+r_{n+1}-r_n] - b_n)$$

which implies the equation

$$M([\mathfrak{U}, I]) = M([\mathfrak{U}, I+q]) .$$

Consequently, by induction,

$$M([\mathfrak{U}, I]) = M([\mathfrak{U}, I+kq]) \qquad (k = 1,2,\ldots)$$

which yields $M([\mathfrak{U},I]) = 0$. Thus M vanishes identically. By Lemma 1 we infer that the measure μ is degenerate, which contradicts the assumption. The Lemma is thus proved.

LEMMA 3. *The measure* M *corresponding to a non-degenerate limit distribution* μ *is of the form*

$$M([\mathfrak{U}, I]) = \gamma(\mathfrak{U}) \int_I e^{-\alpha p}\, dp ,$$

where γ *is a finite measure on the unit sphere of* H *and* α *is a positive constant.*

Proof. Since $r_n \to \infty$ and, by Lemma 2, $r_{n+1} - r_n \to 0$ for every positive number t we can find a subsequence $\{r_{k_n}\}$ such that $r_{k_n} - r_n \to t$. By (8) and (11) for every set $[\mathfrak{U},I]$ we have the formula

$$(13) \qquad M_{k_n}([\mathfrak{U}, I] - b_{k_n}) = k_n \lambda([\mathfrak{U}, I+r_{k_n}])$$

$$= \frac{k_n}{n} \cdot M_n([\mathfrak{U}, I+r_{k_n}-r_n] - b_n) .$$

Assuming that the boundaries $\partial[\mathfrak{U},I]$ and $\partial[\mathfrak{U}, I+t]$ have the M-measure zero and taking into account that M does not vanish identically we get from (9) and (13) the existence of the limit $\lim_{n\to\infty} \frac{k_n}{n} = g(t)$ and the equation

$$(14) \qquad M([\mathfrak{U}, I]) = g(t)\, M([\mathfrak{U}, I+t]) .$$

Further, by a simple reasoning we infer that the last equation holds for all intervals I , all positive numbers t and all Borel subsets

u of the unit sphere of H. Moreover, for every positive t the right-hand side of (14) is finite. Hence it follows that the measure M is finite on the whole space $H - \{0\}$. Therefore for a fixed subset u of the unit sphere of H we may introduce the notation

$$M([\mathfrak{U}, I]) = f(v) - f(u) \quad , \tag{15}$$

where $0 < u < v$ and $I = \langle u,v \rangle$. Then equation (14) can be rewritten in the form

$$f(v) - f(u) = g(t)(f(v+t) - f(u+t)) \tag{16}$$

where u,v and t are positive numbers, f is a bounded non-decreasing non-negative function. Suppose that the interval-function $M([\mathfrak{U}, \cdot])$ does not vanish identically. Then $g(t) > 1$ for every positive number t. In fact in the opposite case $g(t_0) \leq 1$ we would have, by induction according to (16) the inequality

$$f(v) - f(u) \leq f(v+kt) - f(u+kt) \qquad (k = 1,2,\ldots) .$$

But the right-hand side of this inequality tends to 0 when $k \to \infty$. Thus f would be a constant function which would contradict the assumption that $M([\mathfrak{U}, \cdot])$ does not vanish identically.

Given $0 < u_0 < v_0$ with $f(v_0) - f(u_0) > 0$, we have, by (16), for every pair t_1,t_2 of positive numbers

$$f(v_0) - f(u_0) = g(t_1)(f(v_0+t_1) - f(u_0+t_1))$$

$$= g(t_1)\, g(t_2)(f(v_0+t_1+t_2) = f(u_0+t_1+t_2)) .$$

On the other hand

$$f(v_0) - f(u_0) = g(t_1+t_2)(f(v_0+t_1+t_2) - f(u_0+t_1+t_2)) .$$

Consequently,

$$g(t_1+t_2) = g(t_1)\, g(t_2) \qquad (t_1,t_2 > 0)$$

It is well-known that the only solution of the last equation satisfying the condition $g(t) > 1$ is of the form $g(t) = e^{\alpha t}$ where α is a positive constant. Further more, the function f being continuous outside a countable set is, by (16), continuous everywhere. Setting $v = u + t$ into (16) we get the equation

$$f(u+t) - f(u) = e^{\alpha t}(f(u+2t) - f(u+t)) .$$

which implies the inequality

$$f(u+2t) - 2f(u+t) + f(u) \leq 0 .$$

Thus the function f is concave. Consequently, it is absolutely continuous. Setting

$$f(u) = f(o) + \int_0^u h(p)\, dp ,$$

we have, by (16) .

$$\int_u^v h(p)\, dp = e^{\alpha t} \int_{u+t}^{v+t} h(p)\, dp .$$

Hence we get the equation $h(p) = \gamma e^{-\alpha p}$ almost everywhere, γ being a non-negative constant. Thus, by (15),

$$M([\ddot{u}, I]) = \gamma(\ddot{u}) \int_I e^{-\alpha p}\, dp .$$

It is evident that the set function γ is a finite measure on the unit sphere of H . The Lemma is thus proved.

As a corollary of the Lemmata 1 and 3 we get the following Proposition:

PROPOSITION 1. <u>Each limit distribution</u> μ <u>is a shifted compound Poisson distribution</u>

$$\mu = e(M) * \delta_c ,$$

<u>where</u> c <u>is an arbitrary element of</u> H <u>and</u>

$$M([\mathfrak{U}, I]) = \gamma(\mathfrak{U}) \int_I e^{-\alpha p}\, dp \ ,$$

α being a positive constant and γ a finite measure on the unit sphere of H .

PROPOSITION 2. Each probability measure μ of the form

$$\mu = e(M) * \delta_c \ ,$$

where c is an arbitrary element of H and

$$M([\mathfrak{U}, I]) = \gamma(\mathfrak{U}) \int_I e^{-\alpha p}\, dp \ ,$$

α being a positive constant and γ a finite measure on the unit sphere of H is the limit distribution of a sequence (1) .

Proof. Of course, we may assume that the measure M does not vanish identically because in the remaining case the assertion is obvious. Put $\beta_0^{-1} = M(H)$, $\lambda(E) = \beta_0 M(E)$ and $r_n = \alpha^{-1}(\log n + \log \beta_0)$ for sufficiently large n . Using the notations (2), (4), (5) and (6) we get the relation

$$\mu = \lim_{n\to\infty} \nu_n * \delta_c \ .$$

Hence, according to [1], Corollary 6.1, we get the formula

$$\mu = \lim_{n\to\infty} \lambda_n^{*n} * \delta_c$$

which completes the proof.

Acknowledgement: I would like to thank Professor K.Urbanik for his help in preparation of this paper.

References

[1] K.R. Parthasarathy, Probability measures on metric spaces, New York - London, 1967.

WINTER SCHOOL
ON PROBABILITY
Karpacz 1975
Springer's LNM

472

INVARIANT MEASURES FOR PIECEWISE MONOTONIC TRANSFORMATIONS

By Z. S. Kowalski
Wrocław University

0. Summary.

For a class F of piecewise monotonic transformations of $[0,1]$ all invariant and ergodic sets are characterized as well as some sufficient conditions for ergodicity of elements of F are given. A class of operators Q_f, $f \in L_1$, defined on a subset $F(M,q)$ of F and with values in L_1 is considered. $Q_f(\tau)$ gives the density of τ - invariant measure. A necessary and sufficient condition for the continuity of Q_f on a subset of $F(M,q)$ is found. It turns out that the set on which Q_f is continuous for every $f \in L_1$ contains an open and dense subset of $F(M,q)$.

1. Preliminaries.

A starting point for our considerations is a construction of invariant measure given by Lasota and York [1].

Denote by $(L_1, \| \|)$ the space of all Lebesgue integrable functions defined on the interval $[0,1]$. Lebesque measure on $[0,1]$ will be denoted by m.

(a). A transformation $\tau : [0,1] \to [0,1]$ will be called piecewise C^1, if there exists a partition $0 = a_0 < a_1 < \dots < a_q = 1$ of the unit interval such that for each integer i, $i = 1,\dots,q$, the restriction τ_i of τ to the open interval (a_{i-1}, a_i) is a C^1 -

function which can be extended to a function $\bar{\tau}_i$ which is C^1 - function in $[a_{i-1}, a_i]$.

(b) Denote by F the set of all transformations $\tau : [0,1] \to [0,1]$ such that τ is piecewise C^1 , $\inf |\dot{\tau}| > 1$ and each derivative $\dot{\tau}_i$, satisfies the Lipschitz condition.

Given $\tau \in F$, we denote by q_τ the minimal integer q for which there exists a partition $0 = a_0 < \ldots < a_q = 1$ satisfying the requirements of conditions (a) and (b). We define the subset $F(M,q)$ of F : $\tau \in F(M,q) \iff \tau \in F$, $1 + 1/M \leq |\dot{\tau}| \leq M$, $q_\tau \leq q$ and M is the Lipschitz constant for $\dot{\tau}_i$, $i = 1,\ldots,q$. For any $\tau \in F$ we denote by P_τ the <u>Frobenius - Perron operator</u> acting in L_1 , i.e. for $f \in L_1$

$$P_\tau f(x) = \frac{d}{dx} \int_{\tau^{-1}(0,x)} f(s)ds = \sum_{i=1}^{q_\tau} f(\Psi_i(x))\, \sigma_i(x)\, \chi_i$$

where $\Psi_i = \tau_i^{-1}$, $\sigma_i = |\dot{\Psi}_i|$ and χ_i is the characteristic function of the interval $J_i = \tau_i\,[a_{i-1}, a_i]$. The operator P_τ is linear and continuous and satisfies the following conditions:

(c) P_τ is positive, i.e. $f \geq 0$ implies $P_\tau f \geq 0$;

(d) P_τ preserves integrals ;

$$\int_0^1 P_\tau f \, dm = \int_0^1 f \, dm , \qquad f \in L_1$$

(e) $P_{\tau^n} = P_\tau^n$ (τ^n denotes the n-th iterate of τ)

(f) $P_\tau f = f$ for $f \geq 0$ iff the measure $d\mu = f\,dm$ is invariant under τ , that is $\mu(\tau^{-1}(A)) = \mu(A)$ for any measurable A .

A. Lasota and J. Yorke proved the following Theorem (see [1])

THEOREM A. <u>Let</u> $\tau : [0,1] \to [0,1]$ <u>be a piecewise</u> C^2 <u>function such that</u> $\inf |\dot{\tau}| > 1$. <u>Then for any</u> $f \in L_1$ <u>the sequence</u>

$$\frac{1}{n} \sum_{k=0}^{n-1} P_\tau^k f$$

<u>is norm-convergent to a function</u> $f^* \in L_1$, <u>and the limit function</u> f^* <u>has the following properties</u> :

(1) $f^* \geq 0$ <u>for</u> $f \geq 0$

(2) $\int_0^1 f^* dm = \int_0^1 f \, dm$,

(3) $P_\tau f^* = f^*$ and consequently the measure $d\mu^* = f \, dm$ is τ - invariant ,

(4) the function f^* is of bounded variation. Moreover, there exists a constant c independent of the choice of initial f such that the variation of the limiting f^* satisfies the inequality

$$\bigvee_0^1 f^* \leq c \, \|f\| .$$

Remark 1. The analysis of the proof of Theorem A shows that Theorem A remains true if the assumption concerning τ will be replaced by: $\tau \in F$.

For a function f of bounded variation on $[0,1]$ there exist $f(x+)$ for $0 \leq x < 1$ and $f(x-)$ for $0 < x \leq 1$, and the set of discontinuities of f is at most countable. Consider a new function $\bar{f}$ defined as follows: $\bar{f}(x) = f(x-)$ for $x \neq 0$ and $\bar{f}(0) = f(0+)$ for $x = 0$. Obviously $\bigvee \bar{f} \leq \bigvee f$ and for every function f_1 with bounded variation $\int_0^1 |f - f_1| dm = 0$ implies $\bar{f}(x) = \bar{f}_1(x)$ for every $x \in [0,1]$. Now we can define an operator Q_τ which will be used in the sequel: for any $f \in L_1$ we put

(5) $$Q_\tau f = \bar{f}^* .$$

Finally we shall write

$$A \approx B \quad \text{if} \quad m(A \triangle B) = 0 ,$$
$$A \underset{\sim}{\subset} B \quad \text{if} \quad m(A - B) = 0 ,$$
$$f_1 \approx f_2 \quad \text{if} \quad f_1 = f_2 \quad m - \text{a.e.}$$

2. The decomposition Theorem.

The main theorem of this Section will be preceded by two Lemmata.

LEMMA 1. For any transformation τ in F there exists a number $d_\tau > 0$ such that : if a measurable set A is τ-invariant and there exists an open nonempty interval I , $I \underset{\sim}{\subset} A$, then $m(A) \geq d_\tau$.

Proof. Let $s = \inf |\dot{\tau}|$ and let N be the first integer such that $s^N > 2$. Put $\tau_1 = \tau^N$ and let $0 = b_0 < \dots < b_{q\tau_1} = 1$ be the partition of the unit interval corresponding to the transformation τ_1 . Define a number d_τ by putting

$$d_\tau = \min_{1 \leq i \leq q\tau_1} |b_i - b_{i-1}| .$$

If a measurable set A satisfies the assumption of Lemma then, by definition of τ_1 , the set A is also τ_1-invariant. We may assume that the interval I is the longest one having the property formulated in Lemma. We shall show that in I lie at least two points of the set $\{b_1, \dots, b_{q\tau_1 - 1}\}$.

(6) Assume that the set $\{b_1, \dots, b_{q\tau_1-1}\} \cap I$ has exactly one element, say b_i . b_i decomposes the interval I into two intervals I^1 and I^2 . Let I_1^0 denote the longer of the two I^1 and I^2 and any of them if they have the same length. Let $I_1 = \tau_1(I_1^0)$. Then $m(I_1) \geq s^N m(I_1^0) \geq (s^N/2)\, m(I) > m(I)$, and $I_1 \lessapprox A$ by the condition $I \lessapprox A$ and by τ_1-invariance of the set A , hence I_1 is longer then I and has the same properties as I which yields a contradiction.

(7) Assume that the intersection $\{b_1, \dots, b_{q\tau_1-1}\} \cap I$ is empty. If $I_1 = \tau_1(I)$ then $m(I_1) \geq s^N\, m(I)$ however $I_1 \lessapprox A$, which is impossible. Therefore, for some i , $[b_{i-1}, b_i] \subset I$ which implies $m(A) \geq m(I) \geq d_\tau$.

LEMMA 2. <u>If a measurable set</u> A <u>is</u> τ-<u>invariant and</u> $m(A) > 0$ <u>then there exists a nonempty interval</u> I <u>such that</u> $I \lessapprox A$.

<u>Proof</u>. Suppose a set A satisfies the assumptions of Lemma. By Theorem A and definition (5): variations of $Q_\tau f$ and $Q_\tau 1_A$ (1_A - characteristic function of A) are finite. τ-invariance of the set A implies that

$$Q_\tau 1_A = 1_A\, Q_\tau 1 \tag{8}$$

outside a set B of measure zero. Denote by $\bigvee_0^1 f$ the variation of the function f on the set $B^c = [0,1] - B$. Let $f = 1_A$ and let $g = Q_\tau 1$. Then

$$(9) \qquad \infty > \bigvee_0^1 Q\, 1_A \geq \bigvee_\Omega^1 Q\, 1_A = \bigvee_0^1 f\, g$$

$$\geq \sup \sum_{i=1}^{n} |f(x_i)|g(x_i) - g(x_{i-1})| - g(x_{i-1})|f(x_i) - f(x_{i-1})||$$

$$\geq \sup \sum_{i=1}^{n} g(x_i)|f(x_i) - f(x_{i-1})| - \bigvee_0^1 g \; .$$

where sup is taken over $\{x_0, x_1, \ldots, x_n\}$, $x_i \in B^c$. Because $m(A) > 0$, $fg \neq 0$. There exists a point $x \in A$ which is a point of continuity for the function g and a point of density for the set A and such that $g(x) > 0$. Therefore $g > c$ in $[x-\varepsilon, x+\varepsilon]$ for some positive numbers c, ε and the last member of the foregoing inequality is at least

$$c \bigvee_{x-\varepsilon}^{x+\varepsilon} f - \bigvee_0^1 g \; .$$

Now $m(A \cap [x-\varepsilon, x+\varepsilon]) > 0$ together with

$$\bigvee_{x-\varepsilon}^{x+\varepsilon} 1_A < \infty$$

imply that the set $A \cap [x-\varepsilon, x+\varepsilon]$ is equivalent to a finite union of intervals and each of them satisfies the hypothesis of Lemma.

We say that a measurable set A is τ-ergodic if for every measurable set B such that $m(B) > 0$ and $B \subsetneq A$ the condition $\tau^{-1}(B) \approx B$ implies $B \approx A$. The phrase : interval $[0,1]$ is decomposed into the finite sum of sets $A_1, \ldots, A_n$ will mean that $A_i \cap A_j \approx \emptyset$ for $i \neq j$ and $[0,1] \approx \bigcup_{i=1}^{n} A_i$.

THEOREM 1. Let τ be an element of F. Then the interval $[0,1]$ can be decomposed into a finite number of open sets which are τ-invariant and τ-ergodic. The number p_τ of these sets is not greater then $q_\tau - 1$.

Proof. τ-invariant sets form a σ-field of sets. By Lemma 1 and

2 we know that $m(A) \geq d_\tau$ for any τ-invariant set of positive measure, therefore there exists a decomposition of the interval $[0,1]$ into a finite number p_τ of sets which are τ-invariant and τ-ergodic. We denote by D an open set of maximal measure such that $D \lesssim A$ where A is an τ-invariant and τ-ergodic set. We have $m(D) \geq m(I) > 0$ where the interval I is given by Lemma 2. There exists a finite set C such that the sets $\tau(D) - C$ and $\tau^{-1}(D) - C$ are open. The above property of sets $\tau(D)$ and $\tau^{-1}(D)$ and the inclusions $\tau(D) \lesssim A$ and $\tau^{-1}(D) \lesssim A$ imply that $\tau(D) \lesssim D$ and $\tau^{-1}(D) \lesssim D$. Now we can write $D \lesssim \tau^{-1}(\tau(D)) \lesssim \tau^{-1}(D) \lesssim D$ hence $D \approx \tau^{-1}(D)$, and the ergodicity of the set A implies $A \approx D$.

The inequality $p_\tau \leq q_\tau - 1$ results from the simple observation that if D is any open τ-invariant set then the largest open interval I, $I \lesssim D$., contains at least one point of the partition corresponding to τ. Let D and I be as above. Then there is an integer i $(1 \leq i \leq q_\tau)$ such that $a_i \in I$: if not we put $I_1 = \tau(I)$. Then $I_1 \lesssim D$ and $m(I_1) \geq s\, m(I) > m(I)$ hence I_1 is longer then I which yields a contradiction.

COROLLARY 1. *If* $\tau \in F$ *and* $q_\tau = 2$ *then the transformation* τ *is ergodic*.

Let $J_i = \tau(a_{i-1}, a_i)$, $i = 1,\dots,q_\tau$.

THEOREM 2. *If* $\tau \in F$, $\inf |\dot{\tau}| > 2$, *and* $\bigcap_{i=1}^{q_\tau} J_i \neq \emptyset$ *then* τ *is ergodic*.

Proof. Let $(D_1,\dots,D_{p_\tau})$ be the decomposition of the interval $[0,1]$ into open and τ-invariant sets. We know from the proof of Lemma 1 that if $\inf|\dot{\tau}| > 2$ then there is an integer i_1 $(1 \leq i_1 < q_\tau)$ such that $[a_{i_1-1}, a_{i_1}] \lesssim D_1$ what implies

$$J_{i_1} = \tau(a_{i_1-1}, a_{i_1}) \lesssim D_1 .$$

Thus

$$\emptyset \neq \bigcap_{i=1}^{q_\tau} J_i \subset \bigcap_{i=1}^{p_\tau} J_{i_1} \lesssim \bigcap_{l=1}^{p_\tau} D_l$$

Hence we get $p_\tau = 1$ which finishes the proof.

The implication : τ ergodic $\Longrightarrow \bigcap_{i=1}^{q_\tau} J_i \neq \emptyset$ is, generally, not true wich may be easily shown by means of an example.

THEBREM 3. <u>For every</u> $\tau \in F$ <u>and</u> $f \in L_1$

$$Q_\tau f \approx \Big(\sum_{i=1}^{p} \frac{1}{m(D_i)} 1_{D_i} \int_{D_i} f \, dm \Big) Q_\tau 1 , \quad (p = p_\tau) ,$$

<u>where</u> $\{D_1, \dots, D_p\}$, <u>is a decomposition of</u> $[0,1]$ <u>into open</u> τ-<u>invariant and</u> τ-<u>ergodic sets</u>.

<u>Proof</u>. If a measure μ_i is such that $d\mu_i/dm = Q_\tau f_i$, for $f_i = 1_{D_i} f$, $i = 1, \dots, p$, then μ_i is τ_i-invariant and, by the definition, absolutely continuous with respect to the measure m . The transformation $\tau_i = \tau | D_i$ is ergodic. Therefore a measure in D_i which is τ_i-invariant and absolutely continuous with respect to the measure m is unique up to a constant factor. Thus we have that $Q_\tau f_i \approx c Q_\tau 1_{D_i} \approx c 1_{D_i} Q_\tau 1$,

$$\int_0^1 Q_\tau f_i \, dm = \int_0^1 f_i \, dm = \int_{D_i} f \, dm ,$$

and that $\int_0^1 c 1_{D_i} Q_\tau 1 \, dm = c m(D_i)$ what implies that

$$c = \int_{D_i} f dm / m(D_i) \qquad \text{and} \qquad Q_\tau f_i = \Big(\int_{D_i} f dm / m(D_i) \Big) 1_{D_i} Q_\tau 1 .$$

From the linearity of Q_τ the proof of the Theorem follows.

3. <u>The continuity Theorem</u>.

We define in $F(M,q)$ the metric ρ by putting

$$\rho(\tau_1, \tau_2) = \int_0^1 |\tau_1 - \tau_2| \, dm$$

By using a standard technique one can obtain the following

PROPOSITION 1. The metric space $(F(M,q),\rho)$ is complete.

It is easy to see that the topological space $(F(M,q),\rho)$ is compact. For any $f \in L_1$ we define a transformation $P_f : F(M,q) \to L_1$ by putting

$$P_f(\tau) = P_\tau f \tag{15}$$

THEOREM 4. The transformation $P_f(\tau)$ is a continuous function of τ for each $f \in L_1$.

Proof. Consider P_f with f continuous on $[0,1]$ and let τ_n converge to τ. It suffices to show that from any subsequence $\{\tau_{n'}\}$ of $\{\tau_n\}$ one can extract a subsequenceover which P_f is convergent to $P_f(\tau)$. First of all, proceeding similarly as in the proof of Prop. 1 we can extract from $\{\tau_{n'}\}$ a subsequence $\{\tau_n\}$ such that the corresponding a_i^n converge to a_i, and for any positive integer l, τ_i^n, $\dot{\tau}_i^n$, $i \in I$, converge uniformly to τ_i, $\dot{\tau}_i$ respectively over $[a_{i-1} + 1/l, a_i - 1/l]$. Due to the definition of P_τ we have

$$\|P_{\tau_n}(f) - P(f)\| \leq \sum_{i\in I} \int_0^1 C_i^n(x)dx + 2q(\sup_{[0,1]} f) \sum_{i\notin I} A_i^n$$

$$A_i^n = a_i^n - a_{i-1}^n$$

$$C_i^n(x) = |f(\Psi_i^n(x))\, \sigma_i^n(x)\, \chi_i^n(x) - f(\Psi_i(x))\, \sigma_i(x)\, \chi_i(x)|$$

where Ψ_i^n, σ_i^n, χ_i^n correspond to τ_i^n. For any $i \notin I$, $\lim_n A_i^n = 0$. Also $\lim_n \int_0^1 C_i^n(x)\,dx = 0$, due to the uniform convergence of τ_i^n, $\dot{\tau}_i^n$. Now to show the theorem it suffices to apply the standard argument of approximation of $f \in L_1$ by continuous functions.

Define

$$Q_f(\tau) = Q_\tau f = \lim_n \frac{1}{n} \sum_{k=0}^{n-1} P_f^k(\tau)$$

where $P_f^k(\tau) = P_\tau^k(f)$. Due to the identity $P_\tau^k(f) = P_\tau k\ f$, $P_f^k(\tau)$ depends continuously upon τ, therefore Q_τ is of the first class of Baire as a limit of continuous transformations. $F(M,q)$ is separable. Applying Baire Theorem we get:

THEOREM 5. <u>The set of discontinuity points of the transformation</u> Q_f <u>is of the first category</u>.

We call transformation $\tau \in F(M,q)$ a <u>point of continuity</u> of Q iff for every sequence $\tau_n \in F(M,q)$ such that $\tau_n \to \tau$ in metric ρ, $Q_{\tau_n} f \to Q_\tau f$ in L_1 norm for every $f \in L_1$. A straightforward consequence of the last theorem (and of separability of L_1) is

THEOREM 6. <u>The set</u> $H(M,q)$ <u>of all points of continuity of</u> Q <u>is a dense subset of</u> $F(M,q)$.

In what follows it will be convenient to have the following special sort of convergence of transformations.

Let τ_n, $\tau \in F(M,q)$. We write: $\tau_n \overset{*}{\to} \tau$ if $\lim_n q_{\tau_n} = q_\tau$, $\lim_n a_i^n = a_i$, $i = 1,\dots,q_\tau$, and

$$\lim_n \sup_{x \in A_i^n} |\tau_i^n(x) - \tau_i(x)| = 0$$

where $A_i^n = (a_{i-1}^n, a_i^n) \cap (a_{i-1}, a_i)$. If $q_\tau = q$ then $\tau_n \overset{*}{\to} \tau$ is equivalent to $\rho(\tau_n, \tau) \to 0$.

Let $s = \inf |\dot{\tau}|$ and let $k(\tau)$ be the samllest integer such that $s^{k(\tau)} > 2$. Consider the following subset of $F(M,q)$:

$$B(M,q) = \Big\{\tau : \tau \in F(M,q),\ q_\tau = q\ ,\ B_\tau \quad \overline{\tau}^{-1}(\overline{B}_\tau) = 0 \quad \text{for every}$$

$$l\ ,\ 0 < l \leq k(\tau) - 1\Big\}$$

where $B_\tau = \{a_0, a_1, \dots, a_{q_\tau}\}$ and $\overline{\tau}(x) = \sum_{i=1}^{q_\tau} \overline{\tau}(x)$, $\overline{B}_\tau = B_\tau - \{0,1\}$. $\overline{\tau}_i$ denotes the extension of τ_i onto $[a_{i-1}, a_i]$.

Now, we are going to prove several crucial lemmata that will be used in the proof of the subsequent theorems.

LEMMA 4. *If* $\tau \in F(M,q)$ *then*

$$\bigvee_0^1 Q_\tau f \leqslant C(M,\tau)\ \|f\| \tag{18}$$

where $C(M,\tau)$ *is a continuous function of* M , s , $k(\tau)$, $h_\tau = \min_i |a_i - a_{i-1}|$, *and* $\min_{b_i \in B_\tau k(\tau)} |b_i - b_{i-1}| = h_{\tau k(\tau)}$.

Proof. $\tau \in F(M,q)$ implies that the functions $\sigma_i = |\dot{\Psi}_i|$ where $\Psi_i = \tau_i^{-1}$, $i = 1,\dots,q$, are Lipschitzian with constant $s^{-3}M$. We write it shortly $\sigma_i \in \mathrm{Lip}\,(s^{-3}M)$. Denote $\tau_1 = \tau^N$ where $N = k(\tau)$. The derivatives $\dot{\tau}_{1i} \in \mathrm{Lip}\,(NM^{2N-1})$, $i = 1,\dots,q_{\tau_1}$, and $\sigma_{1i} \in \mathrm{Lip}\,(NM^{2N-1}\, s^{-3N})$.

The first step in proving (18) is the following estimate

$$\bigvee_{Ji} (f \circ \Psi_i)\sigma_i = \int_{Ji} |d(f \circ \Psi_i)\sigma_i|$$

$$\leqslant \int_{Ji} |f \circ \Psi_i||d\sigma_i| + \int_{Ji} |d(f \circ \Psi_i)|\sigma_i$$

$$\leqslant k \int_{Ji} |f \circ \Psi_i|\sigma_i dm + s^{-1} \int_{Ji} |d(f \circ \Psi_i)|$$

Then the repetition of arguments used by Lasota and York in the proof of Theorem A in [1] yields (18). (In the last term k denotes $\{$Lipschitz constant of $\sigma_i\}/\min(\sigma_i)$.)

Remark 2. If $\tau_n \overset{*}{\to} \tau$ then there is an integer n_0 such that $k(\tau_n) = k(\tau)$ for $n \geqslant n_0$, and this results simply from the fact that $\tau_n \overset{*}{\to} \tau$ implies $\inf|\dot{\tau}_n| \to \inf|\dot{\tau}|$.

LEMMA 5. *If* $\tau_n \overset{*}{\to} \tau$ *and* $\tau \in B(M,q)$ then $\tau_n^{k(\tau)} \overset{*}{\to} \tau^{k(\tau)}$.

Proof. Consider the partitions $D_n = \bigcup_{l=0}^{N-1} \overline{\tau}_n^{-l}(\overline{B}_{\tau_n})$ and $D = \bigcup_{l=0}^{N-1} \overline{\tau}^{-l}(\overline{B}_\tau)$ corresponding to τ_n^N and τ^N respectively, for $N = k(\tau)$. We denote $u = (i_1,\dots,i_l)$ and $\tau_u^n = \overline{\tau}_{i_1}^n \circ \dots \circ \overline{\tau}_{i_l}^n$ for $0 \leqslant i_j \leqslant N$, $j \leqslant N$.

Step I . There is an integer n_0 such that for $n \geqslant n_0$ and $j \neq 0, q_\tau$ $(\tau_u^n)^{-1}(a_j^n) = \emptyset$ iff $(\overline{\tau}_u)^{-1}(a_j) = \emptyset$.

(i) if $(\tau_u^n)^{-1}(a_j) = \emptyset$ then the distance from a_j to the closed set $\tau_u^n[a_{i_l-1}, a_{i_l}]$ is positive. By the assumption, $\tau_n \xrightarrow{*} \tau$ i.e.

$$(19) \qquad a_i^n \to a_i \ , \quad \sup_{A_n^i} |\tau_i - {}_i^m| \to 0 \quad \text{as} \quad n \to \infty$$

$A_i^n = [a_{i-1}, a_i] \cap [a_{i-1}^n, a_i^n]$ and $|\dot{\tau}_i| \leq M$, $|\dot{\tau}_i^n| \leq M$, $i = 1,\ldots,q_\tau$.

Hence the sets $\tau_u^n([a_{i_l-1}^n, a_{i_l}^n])$ are convergent to the set $\tau_u([a_{i_l-1}, a_{i_l}])$, which implies, in view of $a_j^n \to a_j$, that there is an integer $n_{u,j}^1$ such that for $n \geq n_{u,j}^1$, $a_j^n \notin \tau_u^n([a_{i_l-1}^n, a_{i_l}^n])$ and $(\tau_u^n)^{-1}(a_j^n) = \emptyset$.

Let $n_1 = \max\{n_{u,j}^1\}$

(ii) If $(\tau_u)^{-1}(a_j) \neq \emptyset$ then $a_j \in \operatorname{int} \tau_u([a_{i_l-1}, a_{i_l}])$. Otherwise $a_j = [\bar{\tau}_{i_1} \circ \ldots \circ \bar{\tau}_{i_k}](a_s)$, $k \leq l$, where $s = i_k$ or $s = i_k - 1$. This implies that $\bar{\tau}^{-k}(\bar{B}_\tau) \cap B_\tau \neq \emptyset$ which is a contradiction with assumption $\tau \in B(M,q)$. Now, condition (19) implies that there is an integer $n_{u,j}^2$ such that for $n \geq n_{u,j}^2$, $a_j^n \in \operatorname{int} \tau_u^n([a_{i_l-1}^n, a_{i_l}])$ and $(\tau_u^n)^{-1}(a_j^n) \neq \emptyset$.

Let $n_2 = \max\{n_{u,j}^2\}$. Now we put $n_0 = \max\{n_1, n_2\}$.

Step II. For $d \in D - B_\tau$ there exists exactly one sequence $i_1,\ldots,i_l,j$ with $j \neq 0,q_\tau$ such that $d = [\bar{\tau}_{i_1} \circ \ldots \circ \bar{\tau}_{i_l}]^{-1}(a_j)$. Let $[\bar{\tau}_{i_1} \circ \ldots \circ \bar{\tau}_{i_l}]^{-1}(a_j) = [\bar{\tau}_{j_1} \circ \ldots \circ \bar{\tau}_{j_k}]^{-1}(a_i)$ and let $l \leq k$. Put $\Psi_i = (\bar{\tau}_i)^{-1}$. Then $[\Psi_{i_l} \circ \ldots \circ \Psi_{i_1}](a_j) = [\Psi_{j_k} \circ \ldots \circ \Psi_{j_1}](a_i) = d$.

The above equality implies three possibilities: (a) $i_l = j_k$, (b) $i_l = j_k - 1$, (c) $i_l = j_k + 1$, In case (b), (c) $d = a_s$ with either $s = i_l$ or $s = j_k$. This implies that $d \in B_\tau \cap \bar{\tau}^{-k}(\bar{B}_\tau)$ which yields a contradiction. Hence (a). By an easy induction we get $i_s = j_s$ for $s = 1,\ldots,l$, hence $a_j = [\Psi_{j_{k-l}} \circ \ldots \circ \Psi_{j_1}]$ which implies, due to $\tau \in B(M,q)$, that $k = l$ and $a_j = a_i$.

By (19) and part I of proof we get $(\tau_u^n)^{-1}(a_j^n) \to (\tau_u)^{-1}(a_j)$ as $n \to \infty$. This together with step (II) implies that there is an integer $n_1 \geq n_0$ such that for $n \geq n_1$, $|D_n| = |D| = d + 1$. Let $0 = d_0^n < d_1^n < \ldots < d_d^n = 1$ and $0 = d_0 < d_1 < \ldots < d_d = 1$ where $d_i^n \in D_n$ and $d_i \in D$ for $i = 1,\ldots,d$ and for $n \geq n_1$. Then

$d_i^n \to d_i$ as $n \to \infty$ which together with (19) finishes the proof.

LEMMA 6. <u>If</u> $\tau_n \overset{*}{\to} \tau$ <u>and</u> $\tau \in B(M,q)$ <u>then there exists a constant</u> r <u>such that</u>

$$\bigvee_0^1 Q\tau_n f \leq r \|f\|, \quad f \in L_1 .$$

<u>Proof</u>. Utilizing Lemma 1 we get the estimate $\bigvee_0^1 Q\tau_n f \leq$ $\leq C(M,\tau_n) \|f\|$, where $C(M,\tau_n)$ is a continuous function of $k(\tau_n)$, s_n, $h^n_{k(\tau_n)}$, h_n. The assumptions of Lemma quarantee that $\lim_n h_n = h$, $\lim_n s_n = s$ and $k(\tau_n) = k(\tau)$ for n large enough. By Lemma 5 we get $\lim_n h^n_{k(\tau)} = h^n_{k(\tau)}$, which finishes the proof.

Let $(D_1^n,\dots,D^n_{p_n})$, $p_{\tau_n} = p_n$, $n = 1,2,\dots$, be the sets corresponding to τ_n and given by Theorem 1.

THEOREM 7. <u>Let a sequence</u> τ_n <u>be</u> ρ-<u>convergent to</u> τ, <u>and</u> $\tau \in B(M,q)$, $\lim_n p_n = p$, $p_\tau = p$. <u>Then there is an integer</u> n_0 <u>such that</u> $p_n = p$ <u>for</u> $n \geq n_0$, <u>and every subsequence of the sequence</u> D_i^n <u>has a subsequence convergent to a set</u> D_i <u>which is</u> τ-<u>invariant and</u> τ-<u>ergodic</u>, $i = 1,\dots,p$.

Before the proof we collect several facts that will be used in it. The existence of n_0 results from $\lim p_n = p$, p_n, p integers. Let $D_i^{n_k}$ be a subsequence of the sequence D_i^n, $i = 1,\dots,p$, $n_1 < n_2 < \dots$.

(c) The unit ball in Hilbert space $(L_2, \|\ \|_2)$ is weakly compact.

(d) If x_n is a sequence of elements of Hilbert space weakly convergent to x_0 and $\|x_n\|_2 \to \|x_0\|_2$ then $\|x_n - x_0\|_2 \to 0$.

By (c) there is a subsequence of the sequence $1_{D_i^{n_k}}$ - we shall denote it by $1_{D_i^n}$ - and a function $f_i \in L_2[0,1]$ such that

$$1_{D_i^n} \to f_i \quad \text{weakly in } L_2 \quad (i = 1,\dots,p) . \tag{20}$$

CLAIM 1. <u>The functions</u> f_i <u>are</u> τ-<u>invariant</u> $(i = 1,\dots,p)$.

<u>Proof of Claim</u>. Consider the sets $s_l([0,1])$, $l = 1,2,\dots$ defined

as follows: $g \in s_l([0,1]) \iff g \in C_o([0,1])$ and $g(x) = 0$ for $x \in \bigcup_{i=1}^{q_\tau} [a_i - 1/l\ ,\ a_i + 1/l]$ where $0 = a_o < a_1 < \dots < a_{q_\tau} = 1$ is the partition corresponding to τ . Remember that for any $g \in s_l([0,1])$, $\tau_n \overset{*}{\to} \tau$ implies $g(\tau_n) \to g(\tau)$. Let $h \in L_\infty$. Then for

$$\gamma_n = \int\limits_o (1_{D_i^n} h - f_i h)\, g(\tau_n)\, dm$$

$$|\gamma_n| \leq |\int\limits_o^1 (1_{D_i^n} - f_i) h\, g(\tau) dm| + |\int\limits_o^1 (1_{D_i^n} - f_i)\, h(g(\tau_n) - g(\tau)) dm| \ .$$

However, by (20) $\int_o^1 (1_{D_i^n} - f_i) h\, g(\tau) dm \to 0$ as $n \to \infty$, and the second term is less than

$$\sup |g(\tau_n) - g(\tau)|\, \|h\|_\infty\, (1 + \|f_i\|) \to 0 \quad \text{as} \quad n \to \infty \ .$$

This implies that $\int_o^1 G_n\, g dm \to 0$ as $n \to \infty$, where $G_n = P_{\tau_n}(1_{D_i^n} h - f_i h)$, if we only make use of the equality $\int_o^1 G_n g dm = \gamma_n$, valid for any $g \in s_l([0,1])$, $l = 1,2,\dots$. We finally get, for any $h \in L_\infty$, that

$$(21) \qquad G_n \to 0 \quad \text{weakly in} \quad L_2$$

due to the estimate $\|G_n\|_2 \leq$ const. Therefore, by Lemma 3, $P_{\tau_n} h \to P_\tau h$ and $P_{\tau_n}(f_i\, h) \to P_\tau(f_i\, h)$ in L_1 norm . Thus we have $1_{D_i^n} P_{\tau_n} h \to f_i P_\tau h$ weakly in L_2 and by (21) $f_i\, P_\tau h \approx P_\tau f_i\, h$. Therefore $\int_o^1 f_i(\tau) h\, dm = \int_o^1 f_i\, P_\tau h\, dm = \int_o^1 P_\tau(f_i\, h) dm = \int_o^1 f_i\, h\, dm$ for every $h \in L_\infty$. This establishes the Claim 1 .

<u>Proof of Theorem</u> 7. By Claim 1, Theorem 1, and by Birkhoff Ergodic Theorem

$$f_i = \sum_{j=1}^{p} c_j\, 1_{D_j} \ .$$

However $\tau \in B(M,q)$. Hence the convergence $\tau_n \overset{*}{\to} \tau$ implies, due to

Lemma 6, that there is a constant $c > 0$ such that

$$\bigvee_0^1 Q_{\tau_n} 1_{D_i^n} \leq c \ , \qquad \bigvee_0^1 Q_{\tau_n} 1 \leq c \ .$$

Using Hellys Theorem, we can choose from the sequence $\{\tau_n\}$ a subsequence which will be denoted by the $\{\tau_n\}$ such that $Q_{\tau_n} 1_{D_i^n} \to h_i$, and $Q_{\tau_n} 1 \to g$ in L_1 norm. We have also $Q_{\tau_n} 1_{D_i^n} \approx 1_{D_i^n} Q_{\tau_n} 1$, and $1_{D_i^n} Q_{\tau_n} 1 \to f_i\, g$ weakly in L_2 , which implies that $h_i \approx f_i\, g$, and $1_{D_i^n} Q_{\tau_n} 1 \to f_i\, g$ in L_1 norm. Hence $1_{D_i^n}\, g \to f_i\, g$ in L_1 norm. Without loss of generality we may assume that $1_{D_i^n}\, g \to f_i\, g$ m-a.e. Now

$$\int_0^1 f_i\, g dm = \lim_n \int_0^1 1_{D_i^n} Q_{\tau_n} 1\ dm = \lim_n m(D_i^n) = a_i \ .$$

The number a_i is positive. The convergence $\tau_n \overset{*}{\to} \tau$ implies (see Lemma 5) that $\tau_n^{k(\tau)} \overset{*}{\to} \tau^{k(\tau)}$. Hence by using Remark 2 and Lemma 1 we obtain $m(D_i^n) \geq 1/2\ d_\tau$ for n large enough. Thus we have $a_i \geq 1/2\ d_\tau$. We may conclude that there is an x such that $f_i(x)\, g(x) > 0$ and the sequence $1_{D_i^{n_k}}(x)\, g(x)$ tends to $f_i(x)\, g(x)$. Moreover, there is exactly one D_i such that $x \in D_i$. Thus

$$1_{D_i^{n_k}}(x) \to \sum_{j=1}^{p} c_i\, 1_{D_j}(x)$$

what implies that $1 = c_i$ and $f_i \geq 1_{D_i}$. So we get $1_{D_i^n} 1_{D_i} \to 1_{D_i}$ weakly in L_2 , and by (c) , $1_{D_i^n} 1_{D_i} \to 1_{D_i}$ in L_1. From $p_n = p$, valid for $n \geq n_1$, we infer that the set D_i is unique such that $1_{D_i^n} 1_{D_i} \to 1_{D_i}$ in L_1 norm . The conditions $\Sigma_{i=1}^p m(D_i) = 1$ and $\Sigma_{i=1}^p m(D_i^n) = 1$ imply that $1_{D_i^n} \to 1_{D_i}$ in L_1 norm, which finishes the proof of Theorem.

THEOREM 8. <u>Let</u> $\tau_n \to \tau$ <u>in metric</u> ρ , $\tau \in B(M,q)$. <u>Then</u>

(22) $$Q_{\tau_n} f \to Q_\tau f$$

<u>in</u> L_1 <u>norm</u> <u>for</u> <u>every</u> $f \in L_1$ <u>iff</u> $\lim_n p_{\tau_n} = p_\tau$.

Proof. ($\Longrightarrow$). From given τ's we can pick a subsequence τ_n such that p_n is convergent, and $1_{D_i^n} 1_{D_i} \to 1_{D_i}$ in L_1 for any $i \leq \min\{p, p_1\}$, where $p_1 = \lim_n p_n$. Applying Theorem 3 with $f = 1_{D_i^n} 1_{D_i}$ and multiplying the obtained equality by 1_{D_i} we get by passing to the limit that

$$1_{D_i} Q_\tau 1 \approx Q_\tau 1_{D_i} \approx (m(D_i)/a_i)\, 1_{D_i} Q_\tau 1 \quad \text{where} \quad a_i = \lim_n m(D_i^n) .$$

The relation $\int_0^1 1_{D_i} Q_\tau 1 \, dm = m(D_i) > 0$ allows us to deduce that $m(D_i) = a_i$. Hence by (c) we get $1_{D_i^n} \to 1_{D_i}$ in L_1 , i.e. $m(D_i^n \triangle D_i) \to 0$ as $n \to \infty$, for $i = 1,\ldots,\min(p_1,p)$. Assume $p \leq p_1$. From the last relation and from $\Sigma_{i=1}^p m(D_i) = 1$ we get $\Sigma_{i=1}^p m(D_i^n) \to 1$ which implies that $p_1 = p$. The same results from the assumption $p_1 \leq p$.

($\Longleftarrow$). Now assume $\lim_n p_n = p$, ($p_n = p_{\tau_n}$ and $p = p_\tau$). To show that $Q_{\tau_n} f \to Q_\tau f$ in L_1 norm it suffices to prove that every subsequence of the sequence $Q_{\tau_n} f$ has a subsequence convergent to $Q_\tau f$ in L_1 norm. To do it we first apply Theorem 7 to get a subsequence for which $\lim_n m(D_i^n \triangle D_i) = 0$ for $i = 1,\ldots,p_\tau$. Then applying Helly's Theorem we pick a new subsequence such that $f_n = Q_{\tau_n} f$ converges in L_1 say to g . Hence by Lemma 3 we get $P_{\tau_n} f_n \to P_\tau g$ in L_1 and $P_\tau g \approx g$ by the equality $P_{\tau_n} f_n \approx f_n$. If now $f \geq 0$ then $1_{D_i} g$ is a density of τ_i-invariant measure $\tau_i = \tau \mid D_i$. The uniqueness of the measure which is τ_i-invariant and absolutely continuous with respect to τ_i-ergodic measure m implies that

$$1_{D_i} g = \int_{D_i} f \, dm \, Q_\tau 1_{D_i} .$$

Due to the linearity of P_τ and Q_τ the above formula is also true for $f \in L_1$ arbitrary signed. Now by Theorem 3

$$1_{D_i} \int_{D_i} f \, dm \, Q_\tau 1 = Q_\tau 1_{D_i} f$$

hence

$$Q_{\tau_n}(1_{D_i^n} f) \to Q_\tau (1_{D_i} f)$$

in L_1 norm. Finally the equalities

$$Q_{\tau_n} f \approx \sum_{i=1}^{p} Q_{\tau_n}(1_{D_i^n} f)$$

imply that $Q_{\tau_n} f \to Q_\tau f$ in L_1 and this completes the proof.

COROLLARY 2. *If* τ *is ergodic and* $\tau \in B(M,q)$ *then* $\tau \in H(M,q)$.

Proof. It follows from the proof of Theorem 8 .

Remark 3. It is easy to see that the set of all transformations which are ergodic with respect to m is a nowhere dense subset of $F(M,q)$ whenever $q \geq 3$.

COROLLARY 3. *A transformation* $\tau_0 \in B(M,q)$ *is a point of continuity for* Q *iff the function* $p : F(M,q) \to \{1,\ldots,q-1\}$ *defined by:* $p(\tau) = p$, *is continuous in the point* τ_0 .

Let $C(M,q) = H(M,q) \cap B(M,q)$. By Theorem 5 the set $C(M,q)$ is dense in $F(M,q)$. By Remark 2 and Lemma 5 the set $B(M,q)$ is open and dense in $F(M,q)$ and this together with Corollary 3 implies.

COROLLARY 4. $C(M,q)$ *is an open and dense subset of* $F(M,q)$.

By Corollary 1 and by Corollary 2 we get theinclusion $H(M,2) \supset B(M,2)$.

COROLLARY 5. *Ergodic transformations form an open set in* $B(M,q)$.

LEMMA 7. *If* $\tau \in H(M,q)$ *and* $p_\tau \geq 2$ *then* $q_\tau = q$.

Proof. Suppose that there exists τ belonging to $H(M,q)$ for which $p_\tau \geq 2$ and $q_\tau < q$. Let $(D_1,\ldots,D_{p_\tau})$ be the decomposition of the unit interval corresponding to τ . We define a sequence of transformations τ_n as follows: let $a_i \in D_1$ for $i = 1,\dots,q_\tau - 1$, and let $[a_i - 1/n , a_i] \subset D_1$ for sufficiently large n. Then we put

$$\tau_n(x) = \begin{cases} \tau(x) & \text{if } x \notin [a_i - 1/n , a_i] , \\ \tau(x) + c & \text{otherwise} \end{cases}$$

and the number c is chosen so that $\tau[a_i - 1/n, a_i] + c \subset D_2$. Obviously $\tau_n \in F(M,q)$ and $\tau_n \to \tau$ as $n \to \infty$ as well as $q_{\tau_n} = q_\tau + 1 \leq q$. The sets $\{D_3,\dots,D_{p_\tau}\}$ are τ_n-invariant and τ_n-ergodic. We shall show that the set $D_1 \cup D_2$ is τ_n-ergodic too. Let $A \subsetneq D_1 \cup D_2$ and let $\tau_n^{-1}(A) \approx A$. Then $\tau_n^{-1}(A \cap D_2) \cap D_2 \approx A \cap D_2$, hence $\tau^{-1}(A \cap D_2) \approx A \cap D_2$. This and τ-ergodicity of D_2 imply that either $D_2 \subsetneq A$ or $A \cap D_2 \approx \emptyset$.

(23) If $A \cap D_2 \approx \emptyset$ then $A \subsetneq D_1$ and $\tau(A) \lesssim A$ therefore $\tau^{-1}(A^c) \lesssim A^c$ where $A^c = D_1 - A$. Now the τ-ergodicity of the set D_1 implies $0 = m(\tau^{-k}(A^c) \cap A) \to m(A^c)m(A)/m(D_1)$ for $k \to \infty$ i.e. either $m(A) = 0$ or $A \approx D_1$. However, $\tau_n^{-1}(D_1) \not\approx D_1$ implies $A \approx \emptyset$. If $A \gtrsim D_2$ then the set $B = D_1 \cup D_2 - A \lesssim D_1$ is τ_n-invariant. By (23) we get $B \approx \emptyset$ and $A \approx D_1 \cup D_2$. So at last the set $D_1 \cup D_2$ is τ_n-invariant and τ_n-ergodic. By using Theorem 3 we have

$$Q_{\tau_n} 1_{D_1} \approx (1_{D_1 \cup D_2} / m(D_1 \cup D_2))\, m(D_1)\, Q_{\tau_n} 1$$

and passing to the limit

$$1_{D_1} Q_\tau 1 \approx 1_{D_1 \cup D_2} \frac{m(D_1)}{m(D_1 \cup D_2)} Q_\tau 1 .$$

Hence $m(D_1) = m(D_1 \cup D_2)$ which is impossible and therefore the assumption $q_\tau < q$ is false.

For $\tau \notin B(M,q)$ the conclussions of Theorem 8 and Corollary 2 may not be true even for τ ergodic and $q_\tau = q$, as is shown by the following example: let $\tau(x) = (\sqrt{5} + 1)x/2 \mod (1)$. τ is ergodic and $\tau \notin B(M,2)$ for $M \geq (\sqrt{5} + 1)/2$. Using Theorem 7 [2] and a formula given by W.Parry (see [2]) for the density of invariant measure in case of r - adic transformation

$$h_r(x) = \sum_{x < \tau^n(1)} \frac{1}{r^n}$$

for $r > 1$, we may show that $\tau \notin H(M,2)$.

Acknowledgement: The problem has been suggested to me by Professor A. Lasota. I would like to thank him for his encouraging support and for his valuable suggestions.

References.

[1] A. Lasota and Y. Yorke, On the existence of invariant measures for piecewise monotonic transformations, Transactions of The Amer. Math. Soc. 1973, 481-488.

[2] W. Parry, On the β-expansion of real numbers, Acta Math. Acad. Sci. Hungar. 11(1960), 401-416.

WINTER SCHOOL
ON PROBABILITY
Karpacz 1975
Springer's LNM

472

A LIMIT THEOREM FOR TRIANGULAR ARRAYS OF REPRESENTATIONS OF CANONICAL ANTI-COMMUTATION RELATIONS

By C.Ładogórski
Wrocław University

The basic feature of infinitely divisible laws in the classical probability theory is that they are limit laws for sums of triangular arrays of independent and uniformly asymptotically negligible random variables. An analogous result can be also formulated in the non-commutative probability theory or, more precisely, in the theory of representations of canonical anti-commutation relations. The concept of infinitely divisible distributions of representations of these relations was introduced by D.Mathon and R.F.Streater in [2]. Their main result says that the class of all infinitely divisible distributions of representations coincides with the class of all normal (Gaussian) distributions. The aim of this paper is to prove that the class of all possible limit distributions of triangular arrays of stochastically independent representations coincides with the class of all infinitely divisible ones. Namely, we shall prove that the class of limit distributions consists of normal distributions.

Let H be a complex Hilbert space with the scalar product $(\cdot,\cdot)$ linear in its second argument.

DEFINITION 1. A k - extended representation of the canonical anti-commutation relations (kRCAR) $(k > 0)$ is a map π from H to the set of bounded operators in a Hilbert space K such that

$$\left\{\pi(f),\ \pi(g)\right\}_+ \stackrel{df}{=} \pi(f)\,\pi(g) + \pi(g)\,\pi(f) = 0$$

$$\left\{\pi^*(t),\ \pi(g)\right\}_+ = k(t,g)\ \mathbb{1}$$

where $\mathbb{1}$ is the unit, and $f,g \in H$.

We shall construct the C^* - algebra $\overline{A_k(H)}$, whose representations are in 1-1 correspondence with k-RCARs.

Let us denote by H^* the dual space to H . By the Riesz theorem there is a canonical bijection $C\colon H \to H^*$ which is defined by $\langle Cf,g\rangle = (f,g)$, $g \in H$. Let $H_0 = H \oplus H^*$. We can define on H_0 a conjugation, denoted by $*$, as follows:

$$f^* = Cf$$

$$Cf^* = f \qquad\qquad \text{for } f \in H$$

Let now $P(H)$ denote the polynominal algebra over H_0 . The conjugation on H_0 extends in a unique way to an involution on $P(H)$. Let G_k be the 2-sided $*$ - ideal in $P(H)$ generated by $\left\{\{f,g\}_+,\ \{f^*,g\}_+ - k\langle f^*,g\rangle \mathbb{1};\ f,g \in H\right\}$. The algebra $A_k(H) = P(H)/G_k$ inherits the $*$ - structure from $P(H)$ and it is a well - known fact that there is a unique norm in $A_k(H)$ defining a C^* - algebra $\overline{A_k(H)}$. We shall denote the canonical map $P(H) \to \overline{A_k(H)}$ by a_k .

Representations of $\overline{A_k(H)}$ are, naturally, in 1-1 correspondence with k-RCARs , so we use the same latter π for k-RCAR and corresponding to it representation of $\overline{A_k(H)}$.

For a π-k-RCAR and a ρ - state (i.e. a selfadjoint, positive operator on K with trace equal to 1) the function $\omega_\rho^\pi(x) = \mathrm{tr}\ \rho\pi(x)$ $x \in \overline{A_k(H)}$ is a state on $\overline{A_k(H)}$ i.e. ω_ρ^π has the properties:

(i) ω_ρ^π is linear ,

(ii) $\omega_\rho^\pi(1) = 1$,

(iii) $\omega_\rho^\pi(x^*x) \geq 0$ for all $x \in \overline{A_k(H)}$,

Conversely for every state on $\overline{A_k(H)}$, by Gelfand-Segal construction, there exist a cyclic representation of $\overline{A_k(H)}$ such that $\omega(x) = (\Omega,\pi(x)\ \Omega)$ for all $x \in \overline{A_k(H)}$ where Ω is a cyclic vector and $\|\Omega\| = 1$. So $\omega = \omega^\pi_{P_\Omega}$ where P_Ω is a projector on the subspace generated by vector Ω .

DEFINITION 2. The functional $W_\rho^\pi(x) = \omega_\rho^\pi(a_k(x))$ $(x \in P(H))$ will be called the <u>distribution of a k-RCAR</u> π in a state ρ .

It has the following properties:

(i) W_ρ^π is linear ,

(ii) $W_\rho^\pi(1) = 1$,

(iii) $W_\rho^\pi(x^*x) \geq 0$,

(iv) $W_\rho^\pi(G_k) = 0$.

Conversely, if W satisfies (i), (ii), (iii) and (iv) then W is a distribution of some representation in a certain state.

DEFINITION 3. Function $W_\rho^\pi(F_1 F_2)$, $F_1, F_2 \in H_0$, is called the <u>covariance of</u> π in the state ρ .

DEFINITION 4. A distribution W is said to be <u>even</u> if W vanishes on homogeneous polynomial of odd degree.

Let $\pi_1, \dots, \pi_n$ be k_i-RCARs $i = 1, \dots, n$. Such a sequence is said to be physically independent if its members anti-commute i.e. if

$$\left\{\pi_j(f), \pi_k(g)\right\}_+ = \left\{\pi_j^*(f), \pi_k^*(g)\right\}_+ = \left\{\pi_j^*(f), \pi_k(g)\right\}_+ = 0$$

for $j \neq k$ and $f, g \in H$.

DEFINITION 5. A sequence $\pi_1, \dots, \pi_n$ of RCARs is said to be <u>stochastically independent</u> if it is physically independent and

$$\operatorname{tr} \rho \prod_{j=1}^{k} \pi_j(a_{k_j}(x_j)) = \prod_{j=1}^{k} \operatorname{tr} \rho \pi_j(a_{k_j}(x_j))$$

for $x_j \in P(H)$.

<u>Remark</u> 1. If π_1 and π_2 are stochastically independent then either π_1 or π_2 is even. In fact, if k and s were odd, then

$$\operatorname{tr} \rho\pi_1(f_1^{\sigma_1}) \cdot \dots \cdot \pi_1(f_k^{\sigma_k})\, \pi_2(g_1^{\sigma_1'}) \cdot \dots \cdot \pi_2(g_s^{\sigma_s'})$$

$$= \operatorname{tr} \rho\pi_1(f_1^{\sigma_1}) \cdot \dots \cdot \pi_1(f_k^{\sigma_k}) \operatorname{tr} \rho\pi_2(g_1^{\sigma_1'}) \cdot \dots \cdot \pi_2(g_s^{\sigma_s'}) \quad .$$

and from anticommutativity of π_1 and π_2

$$\operatorname{tr}\rho\pi_1(f_1^{\sigma_1})\cdot\ \dots\ \cdot\pi_1(f_k^{\sigma_k})\,\pi_2(g_1^{\sigma_1'})\cdot\ \dots\ \cdot\pi_2(g_s^{\sigma_s'})$$

$$= -\operatorname{tr}\rho\pi_2(g_1^{\sigma_1'})\cdot\ \dots\ \cdot\pi_2(g_s^{\sigma_s'})\,\pi_1(f_1^{\sigma_1})\cdot\ \dots\ \cdot\pi_1(f_k^{\sigma_k})$$

$$= -\operatorname{tr}\rho\pi_2(g_1^{\sigma_1'})\cdot\ \dots\ \cdot\pi_2(g_s^{\sigma_s'})\operatorname{tr}\rho\pi_1(f_1^{\sigma_1})\cdot\ \dots\ \cdot\pi_1(f_k^{\sigma_k})$$

σ means 0 or 1, and correspondingly f^{σ} means f or f^*. Hence

$$\operatorname{tr}\rho\pi_1(f_1^{\sigma_1})\cdot\ \dots\ \cdot\pi_1(f_k^{\sigma_k})\operatorname{tr}\rho\pi_2(g_1^{\sigma_1'})\cdot\ \dots\ \cdot\pi_2(g_s^{\sigma_s'}) = 0\ .$$

Consequently either π_1 or π_2 is even in a state ρ.

Remark 2. If $\pi_1,\dots,\pi_n$ are stochastically independent then each pair π_j, π_k is stochastically independent hence all are even except for at most one of them.

DEFINITION 6. For an even distribution W we define the truncated functionals (also called cumulants) W_T on $P_1(H)$, the subalgebra of $P(H)$ consisting of all polynomials with no constant term, by induction

$$W_T(F) = 0$$

$$W(F_1\cdot\ \dots\ \cdot F_n)$$

$$= \sum_P (-1)^{\varepsilon(P)} W_T(F_{i_1}\cdot\ \dots\ \cdot F_{i_{l_1}})\cdot\ \dots\ \cdot W_T(F_{i_{l_{k-1}+1}}\cdot\ \dots\ \cdot F_{i_{l_k}}),\ F_i \in H_o,$$

where the sum is taken over all partitions of $(1,\dots,n)$ into disjoint subsets $(i_1,\dots,i_{l_1}),\dots,(i_{l_{k-1}+1},\dots,i_{l_k})$. The indices in each cluster appear in their natural order and $\varepsilon(P)$ is the parity of the permutation $(1,\dots,n) \to (i_1,\dots,i_{l_1},\dots,i_{l_k})$.

W_T is well defined since W_T vanishes on the elements $x \in P_1(H)$ of odd degree and hence the permutations of clusters do not alter $\varepsilon(P)$. Conversely W is uniquely determined by W_T on $P_1(H)$ and the condition $W(1) = 1$.

DEFINITION 7. A distribution W is said to be <u>normal</u> (<u>Gaussian</u>) if it is even and corresponding cumulant W_T vanishes on a homogeneous polynomials of degree > 2.

LEMMA (see [1]). <u>For stochastically independent and even</u> $\pi_1,\dots,\pi_n$ k_i-RCARs , <u>in a state</u> ρ

$$W_{\rho T}^{\sum_{i=1}^{n}\pi_i} = \sum_{i=1}^{n} W_{\rho T}^{\pi_i}$$

<u>Proof</u>. We shall prove the result in the case $n = 2$. The general case is similar but notationally cumbersome. We shall prove that

$$W_{\rho T}^{\pi_1+\pi_2}(F_1\cdot \dots \cdot F_{2n}) = W_{\rho T}^{\pi_1}(F_1\cdot \dots \circ F_{2n}) + W_{\rho T}^{\pi_2}(F_1\cdot \dots \cdot F_{2n}) ,$$

$F_i \in H_0$, by induction on n .
Firstly for $n = 1$

$$W_{\rho T}^{\pi_1+\pi_2}(F_1\ F_2)$$

$$= \operatorname{tr} \rho\pi_1(a_{k_1}(F_1)) + \pi_2(a_{k_2}(F_1)))(\pi_1(a_{k_1}(F_2)) + \pi_2(a_{k_2}(F_2)))$$

$$= \operatorname{tr} \rho\pi_1(a_{k_1}(F_1))\ \pi_1(a_{k_1}(F_2)) + \operatorname{tr} \rho\pi_2(a_{k_2}(F_1))\ \pi_2(a_{k_2}(F_2))$$

$$+ \operatorname{tr} \rho\pi_1(a_{k_1}(F_1))\ \pi_2(a_{k_2}(F_2)) + \operatorname{tr} \rho\pi_2(a_{k_2}(F_1))\ \pi_1(a_{k_1}(F_2))$$

The two last terms vanish because of independence and evenness of π_1 and π_2 so that

$$W_{\rho T}^{\pi_1+\pi_2}(F_1\ F_2) = W_{\rho T}^{\pi_1}(F_1\ F_2) + W_{\rho T}^{\pi_2}(F_1\ F_2) .$$

For general n we have

$$\sum_P (-1)^{\varepsilon(P)} W_{\rho T}^{\pi_1+\pi_2}(I_1)\cdot \ \ldots \ \cdot W_{\rho T}^{\pi_1+\pi_2}(I_q)$$

$$= \operatorname{tr} \rho \prod_{j=1}^{2n} (\pi_1(a_{k_1}(F_j)) + \pi_2(a_{k_2}(F_j))$$

$$= \operatorname{tr} \rho \prod_{j=1}^{2n} (\pi_1(a_{k_1}(F_j)) + \operatorname{tr} \rho \prod_{j=1}^{2n} \pi_2(a_{k_2}(F_j))$$

$$+ \sum_{\{J,J'\}} (-1)^{\varepsilon(\{J,J'\})} \operatorname{tr} \rho \prod_{j\in J} \pi_1(a_{k_1}(F_j)) \operatorname{tr} \rho \prod_{j\in J'} \pi_2(a_{k_2}(F_j))$$

where I_r denotes the monomial $F_{i_{l_{r-1}}+1}\cdot \ \ldots \ \cdot F_{i_{l_r}}$ for $(i_{l_{r-1}}+1,\ldots,i_{l_r})$ element of partition of set $(1,\ldots,2n)$. The sum $\sum_{\{J,J'\}}$ stands for the sum over all partitions of the set $(1,\ldots,2n)$ into two naturally ordered subsets of even cardinality, and $(-1)^{\varepsilon(\{J,J'\})}$ is the sign of the corresponding permutation. Expanding each trace in terms of cumulants the last expression becomes equal to

$$W_{\rho T}^{\pi_1}(F_1\cdot \ \ldots \ \cdot F_{2n}) + W_{\rho T}^{\pi_2}(F_1\cdot \ \ldots \ \cdot F_{2n})$$

$$+ \sum_P{}' (-1)^{\varepsilon(P)} W_{\rho T}^{\pi_1}(I_1)\cdot \ \ldots \ \cdot W_{\rho T}^{\pi_1}(I_g)$$

$$+ \sum_P{}' (-1)^{\varepsilon(P)} W_{\rho T}^{\pi_2}(I_1)\cdot \ \ldots \ \cdot W_{\rho T}^{\pi_2}(I_q)$$

$$+ \sum_{\{J,J'\}} (-1)^{\varepsilon(\{J,J'\})} \sum_{P_J} (-1)^{\varepsilon(P_J)} W_{\rho T}^{\pi_1}(I_{J_1})\cdot \ \ldots$$

$$\times W_{\rho T}^{\pi_1}(I_{J_q}) \sum_{P_{J'}} (-1)^{\varepsilon(P_{J'})} W_{\rho T}^{\pi_2}(I_{J'_q})\cdot \ \ldots \ \cdot W_{\rho T}^{\pi_2}(I_{J'_q}) \quad .$$

Where $\sum_P{}'$ is the sum over all nontrivial partitions. P_J and $P_{J'}$

denote partitions of J and J′ correspondingly. Using the fact that

$$(-1)^{\varepsilon(P)} = (-1)^{\varepsilon(J,J')}(-1)^{\varepsilon(P_J)}(-1)^{\varepsilon(P_{J'})}$$

where P is the partition of (1,...,2n) induced by partitions P_J and $P_{J'}$ we get that the above expression is equal to

$$W_{\rho T}^{\pi_1}(F_1\cdot \ldots \cdot F_{2n}) + W_{\rho T}^{\pi_2}(F_1\cdot \ldots \cdot F_{2n})$$

$$+ \sum_P{}' (-1)^{\varepsilon(P)}(W_{\rho T}^{\pi_1} + W_{\rho T}^{\pi_2})(I_1)\cdot \ldots \cdot (W_{\rho T}^{\pi_1} + W_{\rho T}^{\pi_2})(I_q)$$

and by inductive hypothesis we, finally, obtain that

$$\sum_P (-1)^{\varepsilon(P)} W_{\rho T}^{\pi_1+\pi_2}(I_1)\cdot \ldots \cdot W_{\rho T}^{\pi_1+\pi_2}(I_q)$$

$$= W_T^{\pi_1}(F_1\cdot \ldots \cdot F_{2n}) + W_T^{\pi_2}(F_1\cdot \ldots \cdot F_{2n})$$

$$+ \sum_P{}'(-1)^{\varepsilon(P)} W_{\rho T}^{\pi_1+\pi_2}(I_1)\cdot \ldots \cdot W_{\rho T}^{\pi_1+\pi_2}(I_q) ,$$

so that the result follows by subtracting the second term on the right from both sides.

DEFINITION 8. A sequence of distributons W_n is said to be <u>convergent</u> if $W_n(x)$ converges pointwise for all $x \in P(H)$

<u>Remark</u> 3. If $W_n(x)$ converges pointwise to $W(x)$ where W_n is a distribution of k_n-RCAR then k_n converges to k and $W(x)$ is a distribution. Indeed, it is suffitient to verify conditions (i) to (iv) for the limit $W(x)$. (i), (ii) and (iii) are trivial; (iv) follows from the equalities:

$$W_n(F_1\cdot \ldots \cdot F_i \{f,g\}_+ F_{i+1}\cdot \ldots \cdot F_j) = 0$$

$$W_n(F_1\cdot \ldots \cdot F_i\{f^*,g\}_+ F_{i+1}\cdot \ldots \circ F_j) = k_n\langle f^*,g\rangle W_n(F_1\cdot \ldots \circ F_j)$$

Tending with n to infinity we get that

$$W(F_1\cdot \ldots \cdot F_i\{f,g\}_+ F_{i+1}\cdot \ldots \cdot F_j) = 0$$

$$W(F_1\cdot \ldots \cdot F_i\{f^*,g\}_+ F_{i+1}\cdot \ldots \cdot F_j) = k\langle f^*,g\rangle W(F_1\cdot \ldots \cdot F_j)$$

and at the same time we obtain convergence k_n to k. If W_n are even and $W_n \to W$ then W is also even. For even distribution convergence of W_n is equivalent to convergence of their cumulants W_{nT}.

THEOREM. Let π_{ij} be a triangular array of k_{ij}-RCARs, $i = 1,\ldots,j$ $j = 1,2,3,\ldots$, and ρ a state such that π_{ij} are stochastically independent for $i = 1,\ldots,j$ (i.e. in rows) in a state ρ, $k_{ij} \to 0$ for $j \to \infty$ uniformly with respect to i, and the covariances of sums $\sum_{i=1}^{j} \pi_{ij}$ converges

$$W_\rho^{\sum_{i=1}^{j}\pi_{ij}}(F_1\ F_2) \to W(F_1\ F_2)$$

Then the distributions of sums $\sum_{i=1}^{j} \pi_{ij}$ converge to the normal distribution with covariance equal to $W(F_1\ F_2)$

Proof. First we assume that π_{ij} have even distributions. Let W_{ij} denote the distribution of π_{ij} in a state ρ and W_{ijT} corresponding cumulant.

$$|W_{ij}(f_1^{\sigma_1}\circ \ldots \cdot f_n^{\sigma_n})| = |\mathrm{tr}\ \rho\pi_{ij}(f_1^{\sigma_1})\cdot \ldots \cdot\pi_{ij}(f_n^{\sigma_n})|$$

$$\leq \|\pi_{ij}(f_1^{\sigma_1})\|\cdot \ldots \cdot\|\pi_{ij}(f_n^{\sigma_n}\| \leq k_{ij}^{n/2}\|f_1\|\cdot \ldots \cdot\|f_n\|$$

The last inequality follows from the fact that $\|\pi(f)\| \leq \|f\|$ for 1-RCAR and that if π is k-RCAR then $k^{-1/2}\pi$ is 1-RCAR. Now, we shall show the analogous inequality for W_{ijT}. Namely we prove that

$$|W_{ijT}(f_1^{\sigma_1}\cdot \ldots \cdot f_{2n}^{\sigma_{2n}})| \leq a_n k_{ij}^n \|f_1\| \cdot \ldots \cdot \|f_{2n}\|$$

for some sequence of numbers (a_n) . We proceed by induction. For $n = 1$

$$|W_{ijT}(f_1^{\sigma_1} f_2^{\sigma_2})| = |W_{ij}(f_1^{\sigma_1} f_2^{\sigma_2})| \leq k_{ij}\|f_1\|\|f_2\| .$$

If this inequality holds for $k < n$ then

$$|W_{ijT}(f_1^{\sigma_1}\cdot \ldots \cdot f_{2n}^{\sigma_{2n}})|$$

$$= |W_{ij}(f_1^{\sigma_1}\cdot \ldots \cdot f_{2n}^{\sigma_{2n}}) - \sum_P{}' (-1)^{\varepsilon(P)} W_{ijT}(I_1)\cdot \ldots \cdot W_{ijT}(I_q)|$$

$$\leq |W_{ij}(f_1^{\sigma_1}\cdot \ldots \cdot f_{2n}^{\sigma_{2n}})| + \sum_P{}' |W_{ijT}(I_1)| \cdot \ldots \cdot |W_{ijT}(I_q)|$$

$$\leq k_{ij}^n \|f_1\| \cdot \ldots \cdot \|f_{2n}\| + \sum_P{}' a_{i_1} \cdot \ldots \cdot a_{i_q} k_{ij}^n \|f_1\| \cdot \ldots \cdot \|f_{2n}\|$$

$$= a_n k_{ij}^n \|f_1\| \cdot \ldots \cdot \|f_{2n}\|$$

where $a_n = 1 + \sum_P{}' a_{i_1} \cdot \ldots \cdot a_{i_q}$ and a_{i_r} denotes 1/2 degree of the monomial I_r or 1/2 cardinality of the corresponding element of partition P . From the convergence of covariances

$$W_\rho^{\sum_{i=1}^{j} \pi_{ij}} (F_1 F_2) \to W(F_1 F_2)$$

follows the convergence

$$\sum_{i=1}^{j} k_{ij} \to k .$$

Let us consider now the cumulant of $\sum_{i=1}^{j} \pi_{ij}$,

$$W_{\rho T}^{\sum_{i=1}^{j}\pi_{ij}}(F_1\ F_2) = W_{\rho}^{\sum_{i=1}^{j}\pi_{ij}}(F_1\ F_2) \overset{j\to\infty}{\to} W(F_1\ F_2)$$

$$W_{\rho T}^{\sum_{i=1}^{j}\pi_{ij}}(f_1^{\sigma_1}\cdot \ldots \cdot f_{2n}^{\sigma_{2n}})| = |\sum_{i=1}^{j} W_{ijT}(f_1^{\sigma_1}\cdot \ldots \cdot f_{2n}^{\sigma_{2n}})|$$

$$\leq \sum_{i=1}^{j}|W_{ijT}(f_1^{\sigma_1}\cdot \ldots \cdot f_{2n}^{\sigma_{2n}})| \leq a_n\|f_1\|\cdot \ldots \cdot\|f_{2n}\| \sum_{i=1}^{j} k_{ij}^{n} \ .$$

On other hand

$$\sum_{i=1}^{j} k_{ij}^{n} \to 0$$

for $n > 1$ because

$$\sum_{i=1}^{j} k_{ij}^{n} \leq (\max_{i\leq j} k_{ij})^{n-1} \sum_{i=1}^{j} k_{ij}$$

and

$$\max_{i\leq j} k_{ij} \to 0 \ , \qquad \sum_{i=1}^{j} k_{ij} \to k \ .$$

Hence

$$W_{\rho T}^{\sum_{i=1}^{j}\pi_{ij}}(F_1\cdot \ldots \cdot F_{2n}) \overset{j\to\infty}{\to} 0$$

for arbitrary $F_i \in H_o$ and for $n > 1$ what completes the proof of the Theorem for even RCARs.

If π_{ij} are arbitrary RCARs then (Remark 2) in each row at most one representation is not even. Let us denote it by $\pi_{i_j j}$. Let

$$\pi_j = \sum_{i=1}^{j} \pi_{ij} \ , \qquad \pi_j' = \sum_{\substack{i=1 \\ i\neq i_j}}^{j} \pi_{ij}$$

Then

$$|W_\rho^{\pi_j}(f_1^{\sigma_1}\cdot \dots \cdot f_n^{\sigma_n}) - W_\rho^{\pi_j'}(f_1^{\sigma_1}\cdot \dots \cdot f_n^{\sigma_n})|$$

$$= |\mathrm{tr}\, \rho(\pi_j(f_1^{\sigma_1})\cdot \dots \cdot \pi_j(f_n^{\sigma_n}) - \pi_j'(f_1^{\sigma_1})\cdot \dots \cdot \pi_j'(f_n^{\sigma_n})|$$

$$\leq \|\pi_j(f_1^{\sigma_1})\cdot \dots \cdot \pi_j(f_n^{\sigma_n}) - \pi_j'(f_1^{\sigma_1})\cdot \dots \cdot \pi_j'(f_n^{\sigma_n})\|$$

$$= \|\sum_{s=1}^{n} \pi_j(f_1^{\sigma_1})\cdot \dots \cdot \pi_j(f_{s-1}^{\sigma_{s-1}})(\pi_j-\pi_j')(f_s^{\sigma_s})\, \pi_j'(f_{s+1}^{\sigma_{s+1}})\cdot \dots \cdot \pi_j'(f_n^{\sigma_n})\|$$

$$\leq \|f_1\|\cdot \dots \cdot \|f_n\| \sum_{s=1}^{n} (\sum_{i=1}^{j} k_{ij})^{\frac{s-1}{2}} k_{i_j j}^{1/2} (\sum_{\substack{i=1 \\ i\neq i_j}} k_{ij})^{\frac{n-s}{2}}$$

$$\leq \|f_1\|\cdot \dots \cdot \|f_n\|\, k_{i_j j}\, n\, (\sum_{i=1}^{j} k_{ij})^{\frac{n-1}{2}} .$$

Because

$$k_{i_j j} \overset{j\to\infty}{\to} 0 , \qquad \sum_{i=1}^{j} k_{ij} \to k ,$$

the difference

$$|W_\rho^{\pi_j}(F_1\cdot \dots \cdot F_n) - W_\rho^{\pi_j'}(F_1\cdot \dots \cdot F_n)| \to 0$$

for arbitrary $F_i \in H_o$. Consequently $W_\rho^{\pi_j} \to W$ if, and only if $W_\rho^{\pi_j} \to W$. But π_j' is a sum of even RCARs for which the Theorem has been already established.

COROLLARY. <u>Assume that</u> π_{ij} - <u>form a null array of</u> k_{ij}-RCARs <u>i.e.</u> $W_\rho^{\pi_{ij}}(x) \overset{j\to\infty}{\to} 0$ <u>uniformly with respect to</u> i <u>for each</u> $x \in P(H)$, π_{ij} <u>are stochastically independent in rows and</u>

$$W_\rho^{\sum_{i=1}^{j} \pi_{ij}} \to W .$$

Then W is a normal distribution.

References

[1] R.L. Hudson, A quantum - mechanical central limit theorem for anti-commuting observables, J.Appl. Prob. 10 (1973), 502-509.

[2] D. Mathon and R.F. Streater, Infinitely divisible representations of Clifford algebras, Z.Wahrscheinlichkeitstheorie verw. Geb. 20 (1971), 308-316.

WINTER SCHOOL
ON PROBABILITY
Karpacz 1975
Springer's LNM
472

NON-COMMUTATIVE PROBABILITY THEORY ON VON NEUMANN ALGEBRAS

By M. J. Mączyński
Warsaw Technical University

In [4] S. Gudder and J. P. Marchand developed an extensive probability theory on von Neumann algebras. In this lecture we would like to present an introduction to this theory. One of the basic notions of non-commutative probability is, similarly as in classical case, conditional expectation with respect to a subalgebra. As shown in [4] the properties of this expectation are closely related to Gleason's theorem on measures on the closed subspaces of a Hilbert space. This theorem allows us to express properties of a measure by the properties of the corresponding density operator. In this lecture we would like to show that conditional expectation on von Neumann algebras can also be investigated without making use of Gleason's theorem. Instead of introducing density operators we define the concept of a functional or a measure commuting with an operator and relate the existence of conditional expectation to the commutativity properties of the corresponding measure. Finally we discuss the connection between this notion and Gleason's theorem.

This lecture consists of two parts. The first part discusses Gleason's theorem and its interpretation as a Radon-Nikodym theorem for von Neumann algebras and follows closely exposition in [4]. The second part deals with conditional expectation in von Neumann algebras and is based on [6].

1. Gleason's theorem as a generalized Radon-Nikodym theorem

Let H be a Hilbert space, $\underline{A}$ - a von Neumann algebra of bounded operators on H, $P_{\underline{A}}$ - the set of all orthogonal projections in $\underline{A}$. A measure on $P_{\underline{A}}$ is a non-negative mapping $w : P_{\underline{A}} \to R^+$ such that (1) $w(0) = 0$, (2) $w(\Sigma A_i) = \Sigma w(A_i)$ for every countable sequence of mulually orthogonal projections in $P_{\underline{A}}$. An integral w on $\underline{A}$ is a positive linear functional $w : \underline{A} \to C$ satisfying (1) and (2). If w is an integral on $\underline{A}$, then $L_2(\underline{A},w)$ denotes the Hilbert space of operators generated by the operators in $\underline{A}$ with respect to the inner product $(A_1,A_2) = w(A_1{}^*A_2)$, $L_1(\underline{A},w)$ denotes the Banach space of operators genarated by the operators in $\underline{A}$ with respect to the L_1 - norm $\|A\|_1 = w(|A|) = w((A^*A)^{1/2})$. A. Gleason has shown in [3] that for a separable Hilbert space of dimension greater than 2 all measures and consequently integrals in $\underline{A} = B(H)$ are essentially determined by some operators in A.

THEOREM 1.1. (Gleason) _Let_ H _be a separable Hilbert space of dimension grater than_ 2 _and let_ $\underline{A} = B(H)$ _be the von Neumann algebra of all bounded operators on_ H. _If_ w _is a measure on_ $P_{\underline{A}}$ _then there is a unique positive trace class operator_ W _such that_ $w(A) = \mathrm{Tr}\,(WA)$ _for all_ $A \in P_{\underline{A}}$.

As shown in [4], Gleason's theorem can be extended to obtain the following.

THEOREM 1.2. _Let_ H _be a separable Hilbert space with_ $\dim H > 2$ _and let_ $\underline{A} = B(H)$. _If_ w _is an integral on_ $\underline{A}$ _then there exists a unique positive trace class operator_ W _such that_ $w(A) = \mathrm{Tr}\,(WA)$ _for all_ $A \in \underline{A}$. (W _is called the density operator of_ w.)

Proof. Since w restricted to $P_{\underline{A}}$ is a measure, the existence of W follows from Gleason's theorem. We have $w(A) = \mathrm{Tr}\,(WA)$ for all $A \in P_{\underline{A}}$. If $A = \int \lambda P^A(d\lambda)$ is self-adjoint we have

$$w(A) = \int \lambda w(P^A(d\lambda)) = \int \lambda \mathrm{Tr}(WP^A(d\lambda)) = \mathrm{Tr}(W \int \lambda P^A(d\lambda)) = \mathrm{Tr}(WA).$$

If $A \in \underline{A}$ is arbitrary,

$$w(A) = \frac{1}{2} w(A + A^*) - \frac{1}{2} iw(i(A - A^*)) = \mathrm{Tr}(WA).$$

It is clear that W is unique.

It turns out that Gleason's theorem implies that a version of Randon-Nikodym theorem holds for integrals on the von Neumann algebra $B(H)$. Namely, let us say that an integral w is absolutely continuous with respect to an integral v $(w \prec v)$ if $v(P) = 0$ implies $w(P) = 0$ for $P \in P_{\underline{A}}$. Then we have the following.

THEOREM 1.3. (Gudder and Marchand) <u>Let</u> $\underline{A} = B(H)$ <u>be the von Neumann algebra of all bounded operators on a separable Hilbert space</u> H <u>with</u> $\dim H > 2$ <u>and let</u> $w \prec v$ <u>be integrals on</u> $\underline{A}$. <u>Then there exists a unique operator</u> $dw/dv \in L_1(\underline{A}, v)$ <u>such that</u> $w(A) = v(\frac{dw}{dv} A)$ <u>for all</u> $A \in \underline{A}$.

<u>Proof</u>. By Theorem 1.2. there are unique operators W and V such that $w(A) = \mathrm{Tr}\,(WA)$ and $v(A) = \mathrm{Tr}(VA)$ for all $A \in \underline{A}$. Let E_V be the support of V (i.e. the orthogonal complement of the projection onto the null space of V) .Since $w \prec v$ it follows that $E_V W = W E_V = W$. Since V is one-to-one on $E_V H$, the operator $V^{-1}E_V$ is well defined. Define $\frac{dw}{dv} = V^{-1}E_V W$. Then

$$w(A) = \mathrm{Tr}(WA) = \mathrm{Tr}\,(V \frac{dw}{dv} A) = v(\frac{dw}{dv} A)$$

for all $A \in \underline{A}$.

Gleason's theorem can also be related to the following theorem of Dye [1].

THEOREM 1.4. (Dye) <u>Let</u> $\underline{A}$ <u>be a</u> σ - <u>finite, finite von Neumann algebra and</u> $w \prec v$ <u>integrals on</u> $\underline{A}$. <u>Then there exists an operator</u> $T \in L_2(\underline{A}, v)$ <u>such that</u> $w(A) = v(T^*AT)$ <u>for all</u> $A \in \underline{A}$. TT^* <u>is called the generalized Radon-Nikodym derivative of</u> w <u>with respect to</u> v $(TT^* = \frac{dw}{dv})$.

Let us recall that $\underline{A}$ is σ - finite if every collection of mutually orthogonal projections in $\underline{A}$ is at most countable, A is finite if there exists no partial isometry V in $\underline{A}$ such that $VPV^* = I$ for $P \neq I$. If H is separable with $\dim H = \infty$ then $\underline{A} = B(H)$ is σ - finite but not finite and Dye's theorem cannot be applied. It turns out that Gleason's theorem which does hold in this case implies a version of Dye - Radon-Nikodym theorem for this type of von Neumann algebras (Theorem 1.3). If $2 < \dim H < \infty$ then both Gleason and Dye's theorems are valid for $\underline{A} = B(H)$ and the generalized Radon-Nikodym derivative coincides with the density operator.

THEOREM 1.5. (Gudder and Marchand). *If* $2 < \dim H < \infty$ *then* $W = \frac{dw}{d(Tr)}$.

Proof. The trace Tr is an integral on $\underline{A} = B(H)$ which majorizes all other integrals ($w \prec Tr$ for all integrals w) . Hence by the theorems of Gleason and Dye,

$$w(A) = Tr(WA) = Tr(T^*AT) = Tr(TT^*A) = Tr\,(\frac{dw}{d(Tr)}\,A)$$

for all $A \in \underline{A}$. Since W is unique, this implies

$$W = TT^* = \frac{dw}{d(Tr)} \quad .$$

If $\underline{A}$ is abelian and H separable, then $\underline{A}$ is σ - finite and finite and Dye's theorem can be applied. It turns out that in this case $\underline{A}$ arises as a measure algebra on a classical finite measure space and Dye's theorem reduces to the Radon-Nikodym theorem, $\frac{dw}{dv}$ corresponding to the ordinary Radon-Nikodym derivative (for details see [4]).

2. Conditional expectation on von Neumann algebras

In this section we change slightly the terminology used in the previous section. Namely, by a measure on $P_{\underline{A}}$ we shall understand a non-negative mapping $w : P_{\underline{A}} \to R^+$ such that (1) $w(0) = 0$, (2) $w(\Sigma A_i) = \Sigma w(A_i)$ for every *finite* set of mutually orthogonal projections in $P_{\underline{A}}$. If (2) holds for every countable set of mutually orthogonal projections in $P_{\underline{A}}$, then w is said to be a σ - measure. If V is a Banach space, a V - valued measure on $P_{\underline{A}}$ is a map $w : P_{\underline{A}} \to V$ satisfying (1) and (2). If $\underline{A}$ is a von Neumann algebra, $\underline{A}^*$ denotes the dual of $\underline{A}$ (the space of bounded linear functionals on $\underline{A}$) . $\underline{A}$ is assumed to be endowed with the uniform (norm) topology, $\underline{A}^*$ with the weak * - topology. If $w \in \underline{A}^*$, then w restricted to

$P_{\underline{A}}$ is a complex-valued measure. If w is positive, then w restricted to $P_{\underline{A}}$ is a measure. If $w \in \underline{A}^*$ is positive and restricted to $P_{\underline{A}}$ is a σ - measure, then w is said to be an integral (this coincides with the definition of integral in section 1).

DEFINITION 2.1. Let $A \in \underline{A}$ and $w \in \underline{A}^*$. By Aw we shall denote the functional in $\underline{A}^*$ defined by $Aw(B) = w(AB)$ for all $B \in \underline{A}$. Similarly, wA will denote the functional in $\underline{A}^*$ defined by $wA(B) = w(BA)$ for all $B \in \underline{A}$.

It is clear that for a fixed A the map $B \to w(AB)$ is linear and the functional Aw is bounded because

$$\|Aw\| = \sup_{B \neq 0} \frac{\|w(AB)\|}{\|B\|} \leq \frac{\|w\|\|A\|\|B\|}{\|B\|} = \|w\|\|A\|$$

(similarly for the functional wA).

The maps $(A,w) \mapsto Aw$ and $(w,A) \mapsto wA$ are bilinear maps of $\underline{A} \times \underline{A}^*$ and $\underline{A}^* \times \underline{A}$ into $\underline{A}^*$, respectively. For simplicity we shall call them left and right multiplications of functionals by operators. It is evident that the operation of multiplication defined above is associative, i.e. $(Aw)B = A(wB) = AwB$, $(AB)w = A(Bw)$ and so on. When introducing parantheses, one has to distinguish between the evaluation of a functional in $\underline{A}^*$ at some operator in $\underline{A}$ and the multiplication of that functional by this operator. For example, $wAB = (wA)B$ is a functional in $\underline{A}^*$, whereas $wA(B)$ is a complex number (the value of wA at B). Concluding these remarks let us observe that from the definition of weak $*$ topology in A^* it follows that, for a fixed w, the maps $B \to Bw$ and $B \to wB$ are continuous from A into A^*.

DEFINITION 2.2. Let $w \in \underline{A}^*$ and $A \in \underline{A}$. We say that w commutes with $\underline{A}$ if $wA = Aw$. If $X \subset \underline{A}$, we say that w commutes with X if $Aw = wA$ for all $A \in X$.

If $w \in \underline{A}^*$, from the linearity of our multiplication with respect to both factors it follows that the maps $B \to Bw$ and $B \to wB$ are A^* - valued measures on $P_{\underline{A}}$. There arises a natural question under what conditions the map $\overline{B} \to BwB$ is an A^* - valued measure. The answer is given in the following theorem.

THEOREM 2.3. *Let* $\underline{A}$ *be a von Neumann algebra and* $\underline{B} \subset \underline{A}$ *a von Neumann subalgebra of* $\underline{A}$ *(containing the identity* I*). Let* w *be a functional in* A^* *. Then the following conditions are equivalent:*

(1) *The map* $\bar{w} : B \to BwB$ *is an* $\underline{A}^*$ *- valued measure on* $P_{\underline{B}}$.

(2) w *commutes with* $P_{\underline{B}}$.

(3) w *commutes with* $\underline{B}^{-}$.

Proof. It is obvious that (3) implies (2). Assume that (2) holds. Then for any $B \in P_{\underline{B}}$ we have $\bar{w}(B) = BwB = (Bw)B = (wB)B = wB^2 = wB$. Taking $B = \Sigma B_i$ where B_i is a finite set of mutally orthogonal projections in $P_{\underline{B}}$ we get

$$\bar{w}(\Sigma B_i) = w \Sigma B_i = \Sigma wB_i = \Sigma \bar{w}(B_i)$$

(we have used the fact that the map $B \to wB$ is linear). Thus $\bar{w}$ is an A^* - valued measure on $P_{\underline{B}}$.

Next assume that (1) holds. Let $B \in P_{\underline{B}}$. We have $\bar{w}(B + B\ )$ $= \bar{w}(I) = IwI = I$. On the other hand, since $\bar{w}$ is a measure, we have $\bar{w}(B + B\ ) = \bar{w}(B) + \bar{w}(B\ ) = BwB + (I - B)\, w(I - B) = 2BwB - Bw - wB + w$. Comparing both results we get $2BwB = Bw + wB$. Multiplying on the left and on the right by B and taking into account that $B^2 = B$ we obtain $BwB = wB = Bw$, which shows that w commutes with $P_{\underline{B}}$, the set of all projections in $\underline{B}$. Hence (2) holds.

It remains to show that (2) implies (3). We first show that w commutes with all self-adjoint elements in $\underline{B}$. Let $A \in \underline{B}$ be self--adjoint. By the spectral theorem (see e.g. [8]) we can represent A in the form

$$A = \int \lambda P^A(d\lambda)$$

where P^A is the spectral measure corresponding to A . We have $P^A(E) \in P_{\underline{B}}$ for all Borel sets $E \in B(R)$ (note that $I \in \underline{B}$) . Instead of the spectral measure P^A we can form the spectral family E_λ , $\lambda \in R$, corresponding to A (with $E \in P_{\underline{B}}$ for all $\lambda \in R$) and represent A in the form of a uniform limit

$$A = \mathrm{u} - \lim_{\substack{\omega \to 0 \\ n \to \infty}} \sum_{k=1}^{n} \lambda_k' (E_{\lambda_k} - E_{\lambda_{k-1}})$$

where $\omega = \max|\lambda_k - \lambda_{k-1}|$, $-\|A\| = \lambda_0 < \lambda_1 < \ldots < \lambda_n = +\|A\|$ is a partition of the interval $[-\|A\|, +\|A\|]$, and $\lambda_{k-1} < \lambda_k' < \lambda_k$. Using this representation and taking into account that the maps $B \to Bw$ and $B \to wB$ are linear and continuous on $\underline{A}$, we get

$$wA = w \lim_{n\to\infty} \sum_{k=1}^{n} \lambda_k'(E_{\lambda_k} - E_{\lambda_{k-1}})$$

$$= \lim_{n\to\infty} \sum_{k=1}^{n} \lambda_k'(wE_{\lambda_k} - wE_{\lambda_{k-1}}) = \lim_{n\to\infty} \sum_{k=1}^{n} \lambda_k'(E_{\lambda_k}w - E_{\lambda_{k-1}}w)$$

$$= (\lim_{n\to\infty} \sum_{k=1}^{n} \lambda_k'(E_{\lambda_k} - E_{\lambda_{k-1}}))\, w = Aw .$$

Hence w commutes with all self-adjoint elementa in $\underline{B}$. If $T \in \underline{B}$ is arbitrary, T can be represented as $T = A_1 + iA_2$ with $A_1 = \frac{1}{2}(T + T^*)$, $A_2 = -\frac{1}{2}i(T - T^*)$, where A_1 and A_2 are self-adjoint. Now $A_1w = wA_1$ and $A_2w = wA_2$ imply $Tw = wT$. Hence (3) holds. This ends the proof of Theorem 2.3.

For each $B \in P_{\underline{B}}$ $\overline{w}(B)$ is a functional in A^* which can be evaluated at any $A \in \underline{A}$. Since in the weak $*$ - topology a sequence of functionals f_n in $\underline{A}^*$ is convergent to a functional $f \in \underline{A}^*$ if and only if $\lim_{n\to\infty} f_n(A) = f(A)$ for every $A \in \underline{A}$, Theorem 2.3. implies the following corollary.

COROLLARY 2.4. <u>The following conditions are equivalent</u>:

(1) <u>The map</u> $B \to BwB(A)$ <u>is a complex-valued measure on</u> $P_{\underline{A}}$ <u>for every</u> $A \in \underline{A}$.

(2) w <u>commutes with</u> $\underline{B}$.

We also have

COROLLARY 2.5. <u>If</u> $w \in A^*$ <u>is positive and commutes with</u> $\underline{B}$ <u>and</u> A <u>is a positive operator in</u> $\underline{A}$, <u>then the map</u> $\overline{w}_A : B \mapsto BwB(A)$ <u>is a measure on</u> $P_{\underline{B}}$. <u>If, in addition</u>, w <u>is an integral, then</u> $\overline{w}_A$ <u>is a</u> σ - <u>measure</u>.

<u>Proof</u>. In fact, from Corollary 2.4. it follows that this map is a complex-valued measure on $P_{\underline{B}}$, it remains to show that $\overline{w}_A(B) \geq 0$

for all $B \in \underline{P}_{\underline{B}}$. By definition we have $\overline{w}_A(B) = w(BAB)$. Since $A \geq 0$ implies $B^*AB = BAB \geq 0$ (see [8]), and w is positive, we obtain $\overline{w}_A(B) \geq 0$ for all $B \in P_{\underline{B}}$. Hence $\overline{w}_A$ is a measure.

To prove the last part of the corollary we apply a theorem of Dixmier [9] (see also Sakai [10], th. 1.13.2) which states that for a positive functional w , if w is σ - additive on every countable set of mutually orthogonal projections in a von Neumann algebra $\underline{B}$ (i.e. w preserves least upper bounds of countable sets of mutually orthogonal projections), then $w(\text{l.u.b. } A_\alpha) = \text{l.u.b. } w(A_\alpha)$ for every uniformly bounded increasing sequence A_α of positive elements in A (these conditions are in fact equivalent). Consequently, if w is an integral commuting with $\underline{B}$ and $A \in \underline{A}$ is positive, then denoting

$$A_n = \sum_{i=1}^{n} B_i$$

we have

$$\overline{w}_A\ (\Sigma B_i) = \overline{w}_A(\text{l.u.b. } A_n) = w(A\ \text{l.u.b. } A_n) = w(\text{l.u.b.}(AA_n))$$

$$= \text{l.u.b. } w(AA_n) = \text{l.u.b. } \sum_{i=1}^{n} w(AB_i) = \Sigma\, w(AB_i) = \Sigma\, \overline{w}_A(B_i)$$

for every countable set B_i of mutually orthogonal projections in $\underline{B}$. This follows from the theorem mentioned above, because the sequence A_n is increasing and uniformly bounded and consists of positive elements. Hence $\overline{w}_A$ is a σ - measure on $P_{\underline{B}}$.

The corollary above motivates the following definition.

DEFINITION 2.6. Let $\underline{A}$ be a von Neumann algebra, $\underline{B} \subset \underline{A}$ a von Neumann subalgebra, w a positive functional in $\underline{A}^*$ commuting with $\underline{B}$. We shall say that two positive operators A_1 and A_2 in $\underline{A}$ are $(P_{\underline{B}},w)$ - equivalent if the measures $B \mapsto BwB(A_1)$ and $B \mapsto BwB(A_2)$ coincide on $P_{\underline{B}}$.

There arises a question whether for a positive $A \in \underline{A}$ there is a positive A_o which is $(P_{\underline{B}},w)$ - equivalent to A and which belongs to $\underline{B}$. If such A_o exists it is called the conditional expectation of A with respect to $\underline{B}$ and w and is denoted by $E_w(A|\underline{B})$. We refer the reader to [4], for a thorough discussion of the properties of this concept and for examples showing that in case A is an abelian von Neumann algebra arising from a classical probability space, $E_w(A|\underline{B})$ coincides with the usual conditional expectation of A given $\underline{B}$. Here we would like to show that the existence of $E_w(A|\underline{B})$ depends on the commutativity properties of w with respect to $\underline{B}$. Similarly as in [4] we restrict ourselves to considering the case where $P_{\underline{B}}$ is a Boolean algebra of projections (which implies that $\underline{B}$ is abelian).

Before we state the next theorem, let us introduce the following terminology. Let $\underline{A} = B(H)$ be the von Neumann algebra of all bounded operators on H. We shall say that $w \in \underline{A}^*$ is regular if for every $B \in P_{\underline{A}}$ $w(B) = 0$ implies $BwB = 0$. Later we shall show that Gleason's theorem implies that on a separable Hilbert space every bounded positive functional is regular.

THEOREM 2.7. *Let $\underline{A} = B(H)$ be the von Neumann algebra of all bounded operators on a Hilbert space H and let $\mathcal{B}$ be a separable Boolean algebra of orthogonal projections in $\underline{A}$ (generating an abelian von Neumann subalgebra $\underline{B} \subset \underline{A}$). Let w be a regular integral in $\underline{A}^*$ commuting with $\mathcal{B}$. Then for every positive operator $A \in \underline{A}$ there is a positive operator A_0 in $\underline{A}$ whose all spectral projections belong to $\mathcal{B}$ and which is $(\mathcal{B},w)$ - equivalent to A.*

Proof. By Theorem 2.3. w commutes with $\mathcal{B}$ if and only if w commutes with $\underline{B}$. Hence from Corollary 2.5. we infer that $\bar{w}_A : B \to BwB(A)$ is a σ - measure on $\mathcal{B}$. Since, by assumption, $\mathcal{B}$ is separable, it is countably generated and consequently there is a σ - homomorphism P from the Borel algebra $B(R)$ on the real line (which is countably generated and free, see [7]) onto $\mathcal{B}$. This homomorphism is a spectral measure and hence uniquely determines a self-adjoint operator C in $\underline{A}$. The composition $\bar{w}_A \circ P = v_1$ is a σ - measure on $B(R)$ which is absolutely continuous with respect to the measure $v_2 = w^{\mathcal{B}} \circ P$ ($w^{\mathcal{B}}$ denotes the restriction of w to $\mathcal{B}$). In fact, if $v_2(E) = 0$ for some Borel set E, then $w(P(E)) = 0$. Let $P(E) = B$. Since w is regular, $w(B) = 0$ implies $BwB = 0$. Consequently, $\bar{w}_A(B) = BwB(A) = 0$, and $v_1(E) = \bar{w}_A(P(E)) = \bar{w}_A(B) = 0$. Hence $v_1 < v_2$ and the Radon-Nikodym derivative $f(\lambda) = dv_1/dv_2$ exists and is a bounded Borel-measurable function (see [2]). Let $A_0 = \int \lambda f(\lambda)\, P(d\lambda)$. Since $f(\lambda) \geq 0$, A_0 is a positive self-adjoint operator and by the definition of P the spectral projections of A_0

belong to $\bar{\underline{B}}$. It remains to show that A_0 is $(\bar{\underline{B}},w)$ - equivalent to A . We have for any $B = P(E)$

$$BwB(A_0) = Bw(A_0) = w(BA_0) = w(B \int f(\lambda)P(d\lambda))$$

$$= w(P(E) \int f(\lambda)\, P(d\lambda)) = w(\int_E f(\lambda)\, P(d\lambda)) = \int_E f(\lambda)\, w(P(d\lambda))$$

$$= \int_E \frac{dv_1}{dv_2}\, dv_2 = v_2(E) = \bar{w}_A(P(E)) = \bar{w}_A(B) = BwB(A) .$$

Hence A_0 is $(\bar{\underline{B}},w)$ - equivalent to A . This concludes the proof of Theorem 2.7.

We shall now give a practical criterion how to recognize that an integral $w \in \underline{A}^*$ commutes with an operator $A \in \underline{A}$ or with a von Neumann subalgebra of $\underline{A}$. This criterion follows from Gleason's theorem.

THEOREM 2.8. Let H be a separable Hilbert space with $\dim H > 2$ and let $\underline{A} = B(H)$ be the von Neumann algebra of all bounded linear operators on H . Let $w \in \underline{A}^*$ be an integral on $\underline{A}$ and let W be the density operator of w . Then:

(1) w is regular.

(2) If $WA = AW$ for any $A \in \underline{A}$, then $wA = Aw$.

(3) w commutes with a von Neumann subalgebra $\underline{B} \subset \underline{A}$ if and only if $W \in \underline{B}'$ ($\underline{B}'$ denotes the commutant of $\underline{B}$).

Proof. To prove (1) we have to show that $w(B) = 0$ for any $B \in P_{\underline{B}}$ implies $BwB = 0$. If $w(B) = 0$ then $w(B) = Tr(WB) = Tr(WB^2) = Tr(BWB) = 0$. As W is positive and B self-adjoint, BWB is also positive and consequently $BWB = 0$. Now for every $A \in \underline{A}$ we have $BwB(A) = w(BAB) = Tr(WBAB) = Tr(BWBA) = 0$, i.e. $BwB = 0$. Hence (1) holds.

Next assume that $WA = AW$ for some $A \in \underline{A}$. Then for all $B \in \underline{A}$ we have $wA(B) = w(BA) = Tr(WBA) = Tr(AWB) = Tr(WAB) = w(AB) = Aw(B)$, which implies $wA = Aw$. Hence (2) holds.

Property (2) implies that if $W \in \underline{B}'$ then $wA = Aw$ for all $A \in \underline{B}$, i.e. w commutes with $\underline{B}$. Hence (3) holds one way. Conversely, assume that w commutes with $\underline{B}$. In particular, this implies that $wB = Bw$ for any projection $B \in P_{\underline{B}}$. Multiplying this identity by B on the left and on the right we get $BwB = wB$ and $BwB = Bw$. Adding side by side we get $2BwB = wB + Bw$. This implies $2w(BAB) = w(AB) + w(BA)$ for all $A \in \underline{A}$, and consequently $2Tr(WBAB) = Tr(WAB) + Tr(WBA)$, or $2Tr(BWBA) = Tr((BW + WB)A)$. Taking $A = P_\varphi$ the one-dimensional projection onto the unit vector $\varphi \in H$ we infer that $(2BWB\varphi, \varphi) = ((BW + WB)\varphi, \varphi)$ for all $\varphi \in H$, $\|\varphi\| = 1$. Since BWB and $WB + BW$ are self-adjoint, two quadratic forms coincide on the unit sphere of H, which implies that $2BWB = BW + WB$ for all $B \in P_{\underline{B}}$. Multiplying on the left and on the right by B we obtain $BWB = WB = BW$, which shows that W commutes with all projections in $\underline{B}$. Reasoning analogously as in the proof of Theorem 2.3. we show that W commutes with all members of B, i.e. $W \in \underline{B}'$. Hence (3) holds. This ends the proof of Theorem 2.8.

References

[1] Dye, H. A., The Radon-Nikodym theorem for finite rings operators, Trans. Amer. Math. Soc., 72(1952), 243-280.

[2] Halmos, P. R., Measure Theory, Van Nostrand, New York 1950.

[3] Gleason, A. M., Measures on the closed subspaces of a Hilbert space, Journal of Mathematics and Mechanics 6(1957), 885-894.

[4] Gudder, S., and J.-P. Marchand, Non-commutative probability on von Neumann algebras, J. Math. Physics 13(1972), 799-806.

[5] Kato, T., Perturbation Theory for Linear Operators, Springer-Verlag, Berlin-Heidelberg-New York 1966.

[6] Mączyński, M., Conditional expectation in von Neumann algebras, submitted for publication in Studia Mathematica.

[7] Ramsay, A., A theorem on two commuting observables, Journal of Mathematics and Mechanics, 15(1966), 227-234.

[8] Topping, T.M., Lectures on von Neumann Algebras, London 1971.

[9] Dixmier, J., Formes linéaires sur un anneau d'operateurs, Bull. Soc. Math. France, 81(1953), 9-30.

[10] Sakai, S., C^* - algebras and W^* - algebras, Springer-Verlag, Berlin-Heidelberg-New York 1973.

[illegible] on the left and on the right we get [illegible] and [illegible] by [illegible] [illegible] [illegible] (Ma)) [illegible] w(Ma)) [illegible] quantity [illegible] [illegible] [illegible] [illegible] [illegible] dimensional [illegible] of the [illegible] E [illegible] [illegible] [illegible] and [illegible] are [illegible] by quadratic [illegible] on [illegible] [illegible] [illegible] which [illegible] the [illegible] of Theorem [illegible]

References

[1] [illegible], [illegible], The Radon-Nikodym [illegible] Trans. Amer. Math. Soc. [illegible]

[2] [illegible], [illegible] Measure theory, [illegible] New York 1950.

[3] Gleason, [illegible], Measures on the closed subspaces of a Hilbert space, Journal of Mathematics and Mechanics [illegible]

[4] Gudder, S. [illegible] Marchand, [illegible] probability on von Neumann algebras, [illegible]

[5] Kato, T., Perturbation Theory for [illegible] Operators, Springer-Verlag, Berlin-Heidelberg-New York 1966.

[6] [illegible] and [illegible] Neumann [illegible] submitted for publication [illegible]

[7] [illegible], [illegible] of Mathematics and Mechanics [illegible]

[8] [illegible], [illegible] von Neumann Algebras, London [illegible]

[9] [illegible] Math. [illegible] (1953), [illegible]

[10] Sakai, S., C*-algebras and W*-algebras, Springer-Verlag, Berlin-Heidelberg-New York 1971.

WINTER SCHOOL
ON PROBABILITY
Karpacz 1975
Springer's LNM

472

ON UNCONDITIONAL BASES AND RADEMACHER AVERAGES (*)

By A. Pełczyński
Institute of Mathematics, Polish Academy of Sciences

Let (r_j) be the Rademacher orthogonal system i.e.

$$r_j(t) = \text{sign} \sin 2^{(j-1)} 2\pi t , \quad t \in \{0,1\}$$

$j = 1,2,\ldots$. The classical Khinchine inequality says that for each $p \in [1,\infty)$ there are constants A_p and B_p such that

$$(1) \qquad B_p(\Sigma |a_j|^2)^{\frac{1}{2}} \leq (\int_0^1 |\sum_{j=1}^n a_j r_j(t)|^p \, dt)^{\frac{1}{p}} \leq A_p(\Sigma |a_j|^2)^{\frac{1}{2}}$$

for all scalars $a_1,\ldots,a_n$ $(n = 1,2,\ldots)$.

One can ask for what Banach spaces X the inequality (1) is still true if we replace the scalars $a_1,a_2,\ldots$ by elements of X and their absolute values by their norms. Of course, now A_p and B_p depend on X. Kwapień [6] gave, among other results, a complete answer to this question:

THEOREM 1. *Let X be a Banach space; then the following conditions are equivalent*

(*) This lecture was also delivered on the Regional Conference of AMS On the Theory of Best Approximation and Functional Analysis, Kent State University June 11-15, 1973.

(a) <u>for some</u> $p \in [1,\infty)$ <u>there are constants</u> A_p <u>and</u> B_p <u>such that</u>

$$B_p(\Sigma \|a_j\|^2)^{\frac{1}{2}} \leq (\int_0^1 \| \sum_{j=1}^n a_j r_j(t)\|^p) dt^{\frac{1}{p}} \leq A_p(\Sigma \|a_j\|^2)^{\frac{1}{2}}$$

<u>for all</u> $a_1,\ldots,a_n \in X$ $(n = 1,2,\ldots)$;

(b) <u>the same for every</u> $p \in [1,\infty)$;

(c) X <u>is isomorphic to a Hilbert space</u>.

Thus, in the classical form, the Khinchine inequality can only be extended to a very narrow class of Banach spaces. However there is a natural generalization to all Banach spaces. For this observe that, by the orthogonality of the rademacher functions, we have

$$(\sum_{j=1}^n |a_j|^2)^{\frac{1}{2}} = \int_0^1 | \sum_{j=1}^n a_j r_j(t)|^2 dt$$

for all scalars $a_1,a_2,\ldots,a_n$ $(n = 1,2,\ldots)$.
Now, we write the Khinchine inequality in the form

$$(2) \quad B_p(\int_0^1 | \sum_{j=1}^n a_j r_j(t)|^2 dt)^{\frac{1}{2}} \leq (\int_0^1 | \sum_{j=1}^n a_j r_j(t)|^p dt)^{\frac{1}{p}} \leq$$

$$\leq A_p \int_0^1 |\Sigma a_j r_j(t)|^2 dt)^{\frac{1}{2}} \quad .$$

In this new form the inequality is valid for all Banach spaces. Namely we have the Khinchine-Kahane inequality:

THEOREM 2 (cf. Kahane [5]). <u>For every</u> $p \in [1,\infty)$ <u>there exist universal constants</u> α_p <u>and</u> β_p <u>such that, for every Banach space</u> X

$$(3) \quad \beta_p(\int_0^1 \| \sum_{j=1} a_j r_j(t)\|^2 dt)^{\frac{1}{2}} \leq (\int_0^1 \| \Sigma a_j r_j(t)\|^p dt)^{\frac{1}{p}}$$

$$\leq \alpha_p(\int_0^1 \| \Sigma a_j r_j(t)\|^2 dt)^{\frac{1}{2}}$$

<u>for all</u> $a_1, a_2, \dots, a_n \in X$, $n = 1,2,\dots$.

If $(a_j)_{1 \leq j \leq n}$ is a finite sequence of elements of a Banach space X , then the quantity

$$\left(\int_0^1 \| \Sigma\, a_j r_j(t) \|^p dt \right)^{\frac{1}{p}}$$

is called the p-th Rademacher average of the sequence (a_j). The Khinchine-Kahane inequality (3), roughly speaking, says that if we are only concerned with isomorphic properties of a given Banach space X it does not matter what p we choose, i.e. it does not matter whether we deal with 1-Rademacher averages, 2-Rademacher averages or p-th Rademacher averages for a given particular p . In particular it follows immediately from the Khinchine-Kahane inequality that the conditions a) and b) in Theorem 1 are equivalent.

Let us return to the original Khinchine inequality. In view of Theorem 1 one can ask what are the Banach spaces for which one part of the Khinchine inequality remains true. This leads to the following

DEFINITION. A Banach space X is said to have <u>subquadratic</u> (<u>superquadratic</u>) <u>Rademacher</u> <u>averages</u> with constant A_p (resp. B_p) if there exists for some p , equivalently for all p , a constant A_p (resp. B_p) such that for all $a_1, a_2, \dots, a_n \in X$ $(n = 1,2,\dots)$,

$$\left(\int_0^1 \| \sum_{j=1}^n a_j r_j(t) \|^p \right)^{\frac{1}{p}} dt \leq A_p \left(\sum_{j=1}^n \|a_j\|^2 \right)^{\frac{1}{2}}$$

(resp.)

$$B_p \left(\sum_{j=1} \|a_j\|^2 \right)^{\frac{1}{2}} \leq \left(\int_0^1 \| \sum_{j=1}^n r_j(t) \|^p \, dt \right)^{\frac{1}{p}})$$

We shall write $X < R$ whenever X has subquadratic Rademacher averages and $X > R$ whenever X has superquadratic Rademacher averages.

The French [13] use the terminology "X is of type 2" for $X < R$ and "X is of cotype 2" for $X > R$. Let us observe that obviously the properties $X > R$ (resp. $X < R$) are local properties i.e. depend on the metric structure of finite dimensional subspaces of the space X only. Hence by the Local Reflexivity Principle of Lindenstrauss and Rosenthal [8] we have in particular

COROLLARY. *For every Banach space* X , $X > R$ *iff* $X^{**} > R$ ($X < R$ *iff* $X^{**} < R$) .

COROLLARY. *If* $X > R$ *and* Y *is a subspace of* X , *then* $Y > R$ ($X < R$, *then* $Y < R$). *Moreover if* $X < R$ *and* Z *is a quotient of* X , *then* $Z < R$.

Remark. The second part is false for $X > R$ (cf.[1]).

Banach spaces with subquadratic averages behave very much like subspaces of $\mathcal{L}_p$ spaces for $2 \leq p < \infty$ while Banach spaces with superquadratic Rademacher averages behave like $\mathcal{L}_p$ spaces for $1 \leq p \leq 2$. In fact we have.

THEOREM 3. *If* $1 \leq p \leq 2$, *then every subspace of an* $\mathcal{L}_p$ *space has superquadratic Rademacher averages.*

If $2 \leq p < \infty$, *then every subspace of an* $\mathcal{L}_p$ *space has subquadratic Rademacher averages; more generally, this is true for every quotient of a subspace of* $\mathcal{L}_p$.

Theorem 3 is due to Nordlander [10] but in fact is implicitly contained in Orlicz [11]. Of course Nordlander stated the theorem as an inequality for functions in L_p .

Let us also observe that every infinite dimensional $\mathcal{L}_\infty$ space in particular a $C(K)$ space) has neither subquadratic nor superquadratic Rademacher averages. This follows from the fact that if X is an infinite $\mathcal{L}_\infty$ - space, then for every $\varepsilon > 0$ and every finite dimensional space E there is an embedding $T: E \to X$ with $\|T\| \, \|T^{-1}\| < 1 + \varepsilon$.

The class of Banach spaces X with superquadratic (subquadratic) Rademacher averages is much larger than the corresponding classes of subspaces of $\mathcal{L}_p$ spaces. It is closed under the operation of taking products in the sense of l_r for $1 \leq r \leq 2$ (resp. $2 \leq r < \infty$) in the following sense

THEOREM 4. *Let* $1 \leq r \leq 2$ (*resp.* $2 \leq r < \infty$) . *Suppose that* $(X_\gamma)_{\gamma \in \Gamma}$ *is a family of Banach spaces which have superquadratic* (*subquadratic*) *Rademacher averages, each with a constant* $B_{p',\gamma}$ (*resp.*

$A_{p,\gamma}$) , *and suppose that* $\sup_{\gamma} A_{p,\gamma} = A_p < \infty$ (*resp.* $\sup_{\gamma} B_{p,\gamma} = B_p < \infty$). *Then the product* $(\mathbb{P}\, X_2)_r$ *has superquadratic* (*resp. subquadratic*) *Rademacher average.*

On the other hand if $r \neq p$ *the* l_r *product of a family of subspaces of an* $\mathcal{L}_p$ *space is not in general isomorphic neither to a subspace (nor to a quotient, for* $q > 1$) *of an* $\mathcal{L}_q$ *space.* (*cf.* [1] [12]).

There is however a reasonable characterization of those Banach spaces with superquadratic (subquadratic) Rademacher averages which have an unconditional basis or more generally a local unconditional structure.

To state this result I shall need one more concept.

A Banach space X is said to have *the Grothendieck property* if every bounded operator $u: c_0 \to X$ admits a factorization through l_2

$$\begin{array}{ccc} c_0 & \xrightarrow{u} & X \\ \varphi \searrow & & \nearrow \psi \\ & l_2 & \end{array}$$

In the language of absolutely summing operators the Grothendieck property states

"every bounded linear operator from c_0 into X is 2-absolutely summing".

The Grothendieck property is closely related to the superquadraticity and subquadraticity of the Rademacher averages. It is reasonable to make the following

Conjecture: A Banach space X has superquadratic Rademacher averages if and only if it has the Grothendieck property.

B. Maurey proved recently (cf. [9])

THEOREM 5. *If* X *has superquadratic Rademacher averages, then* X *has the Grothendieck property.*

Let me sketch his argument

Assume that X fails to have the Grothendieck property, then by a duality argument there exists an operator

$$w : X^{*} \to l_1$$

which is not factorable through any Hilbert space.

Consider separately two cases

1^{0} For some $p \in (1,2)$ there is a factorization

$$\begin{array}{ccc} X^{*} & \xrightarrow{w} & L_1 \\ & \varphi \searrow \quad \nearrow \psi & \\ & L_p & \end{array}$$

Then $\varphi^{*} : L_{p^{*}} \to X^{**}$ is an operator from the space $L_{p^{*}} < R$ (because $p^{*} \in (2,\infty)$) to the space $X^{**} > R$ (because $X > R$). Now we use the following generalization of Kwapień's Theorem 1.

THEOREM 1 a (Kwapień-Maurey [1] [9]). *If* $X < R$ *and* $Y > R$ *then every bounded operator from* X *into* Y *admits a factorization through a Hilbert space.*

Thus φ^{*} admits such a factorization, and, by duality, φ also does. But this implies that w has a factorization through a Hilbert space, a contradiction.

2^{0} $w : X^{*} \to l_1$ does not factor through any L_p with $1<p<2$. Then we use the recent Maurey-Rosenthal result

(M-R) If $u: E \to L_1$ is a bounded linear operator which does not factor through L_p for some $p > 1$, then for any n there exist elements $e_1^{*}, e_2^{*}, \dots, e_n^{*}$ in E^{*} with $\|e_j^{*}\| = 1$ $(j = 1,2,\dots,n)$ such that

$$\Big\|\sum_{j=1}^{n} t_j e_j^{*}\Big\| \leq 2\Big(\sum_{j=1}^{n} |t_j|^{p^{*}}\Big)^{\frac{1}{p^{*}}}$$

for all scalars $t_1, t_2, \dots, t_n$.

(Rosenthal [12] proved this result for isomorphic embeddings and Maurey [9] generalized it to arbitrary operators).

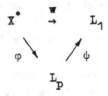

Thus, in our case, for each n there are elements $x_1^{**}, x_2^{**}, \ldots, x_n^{**}$ of norm one in X^{**} such that

$$\|\Sigma\, t_j x_j^{**}\| < 2(\Sigma\, |t_j|^{p^*})^{\frac{1}{p^*}}$$

Putting $t_j = r_j(s)$ and integrating against the Rademacher functions we get

$$(+) \qquad \int_0^1 \|\Sigma_j\, x_j^{**} r_j(s)\| ds < 2\, n^{\frac{1}{p^*}}$$

But $X^{**} > R$, so

$$(++) \qquad \int_0^1 \|\Sigma_j\, x_j^{**} r_j(s)\| ds \geq B_p(\Sigma\, \|x_j^{**}\|^2)^{\frac{1}{2}} \geq B_p\, n^{\frac{1}{2}} .$$

Thus for all n we get $B_p\, n^{\frac{1}{2}} < 2\, n^{\frac{1}{p^*}}$, which is impossible because $p^* > 2$, a contradiction.

We do not know in general whether the Grothendieck property implies the superquadraticity of Rademacher averages. We have, however,

THEOREM 6. _If a Banach space_ X _has an unconditional basis (or more generally a local unconditional structure) then_ X _has the Grothendieck property iff_ X _has superquadratic Rademacher averages_.

Let me sketch the argument for the implication "X has the Grothendieck property" implies "$X > R$" ; assuming that X has an unconditional basis (e_n). It was observed in [1] that if (e_n) is an unconditional basis for X with coefficient functionals (e_n^*) then X has the Grothendieck property

(*) there exists $C > 0$ such that

$$\|\Sigma_n \left(\sum_{j=1}^{m} |e_n^*(x_j)|^2 \right)^{\frac{1}{2}} e_n\| \geq C\, (\Sigma\, \|x_j\|^2)^{\frac{1}{2}}$$

for all $x_1, x_2, \ldots x_m \in X$ and $m = 1, 2, \ldots$.

We show that (*) implies that $X > R$. Fix $x_1, x_2, \ldots, x_m \in X$. Since the basis (e_n) in unconditional, there exists a constant $A > 0$ such that

$$\|\sum_{j=1}^{m} x_j r_j(t)\| = \|\sum_{n} (\sum_{j=1}^{m} e_n^*(x_j)\, r_j(t)) e_n\|$$

$$\geq A\|\sum_{n} |\sum_{j=1}^{m} e_n^*(x_j)\, r_j(t)|\, e_n\|$$

Now, integrating against the Rademacher functions we get, using the Khinchine inequality, the inequality (*), and the assumption that the basis (e_n) is unconditional, the following

$$\int_0^1 \|\sum_{j=1}^{m} x_j r_j(t)\| dt \geq A \int_0 \|\sum_{n} |\sum_{j=1}^{m} e_n^*(x_j) r_j(t)|\, e_n\|$$

$$\geq A\|\sum_{n} \int_0^1 |\sum_{j=1}^{m} e_n^*(x_j) r_j(t)|\, e_n\| \geq A^2 \|\sum_{n} (\sum_{j=1}^{m} |e_n^*(x_j)|^2)^{\frac{1}{2}} e_n\|$$

$$\geq A^2 C \,(\sum_{j=1}^{m} \|x_j\|^2)^{\frac{1}{2}} \quad .$$

Thus $X > R$.
A similar argument combined with recent work of Johnson [4] gives.

THEOREM 7. <u>A Banach space X with an unconditional basis or more generally with a local unconditional structure has subquadratic Rademacher averages if and only if X^* has the Grothendieck property and does not contain subspaces uniformly isomorphic to l_1^n</u> .

There are Banach spaces with superquadratic (subquadratic) Rademacher averages which do not have any local unconditional structure.

Let S_p $(1 \leq p < \infty)$ be the space of all compact operators $T: l_2 \to l_2$ such that $\sigma_p(T) = (\sum_0 |s_j(T)|^p)^{1/p} < \infty$ where $(s_j(T))$ is the sequence of all eigenvalues of the operator $(T^*T)^{1/2}$, each eigenvalue repeated accordingly to its multiplicity. The space S_p, under the norm $\sigma_p(\cdot)$, is a Banach space.

Jaegerman-Tomczak proved recently [14].

THEOREM 8. <u>If</u> $1 \leq p \leq 2$, <u>then</u> $S_p > R$. <u>If</u> $2 \leq p < \infty$, <u>then</u> $S_p < R$.

On the other hand it follows from the recent result of Gordon and Lewis [2] that for $p \neq 2$ the spaces S_p do not have local unconditional structures.

Finally I would like to discuss a probabilistic aspect of Rademacher averages. The Rademacher system (r_j) is a realization of a sequence (δ_n) of independent symmetric equidistributed random variables each distributed according to the law

$$P(\delta_i = 1) = P(\delta_i = -1) = \frac{1}{2} \qquad i = 1,2,\dots$$

The p-th Rademacher average is actually

$$(E \| \sum_{j=1}^{n} x_j \delta_j \|^p)^{\frac{1}{p}} ,$$

the p-th root of the expected value of the random variable $\| \sum_{j=1}^{n} x_j \delta_j \|^p$.

Now we can generalize the concept of Rademacher averages as follows. Let (ξ_n) be any sequence of independent real valued symmetric random variables on a probability space (Ω, Σ, P) . Assume that

$$E\|\xi_n\|^p = \int_\Omega |\xi_n(\omega)|^p \, d_p < +\infty \qquad \text{for } n = 1,2,\dots$$

for some $p \in (0,\infty)$.

A Banach space X is said to have superquadratic (subquadratic) (ξ_n) averages if there exists a constant B_p (resp A_p) such that

$$(E \| \sum_{j=1}^{n} x_j \xi_j \|^p)^{\frac{1}{p}} \geq B_p (\sum_j \|x_j\|^2)^{\frac{1}{2}}$$

$$(\text{resp. } (E \| \sum_{j=1}^{n} x_j \xi_j \|^p)^{\frac{1}{p}} \leq A_p (\sum_j \|x_j\|^2)^{\frac{1}{2}} \ )$$

It turns out that for large class of sequences (ξ_n) we get the same class of spaces as for the Rademacher functions. This follows from recent results of Hoffman-Jorgensen [3]. A particular case of his recent result can be stated as follows.

THEOREM 9. Let (γ_n) be the sequence of independent standard Gaussian variables i.e. for each γ_n

$$P(\gamma_n < t) = \frac{1}{\sqrt{2\pi}} \int_{-\infty}^{t} e^{-\frac{s^2}{2}} ds \quad .$$

Let $1 \leq p < \infty$. Then a Banach space X has a superquadratic (subquadratic) Rademacher averages if and only if X has superquadratic (subquadratic) Gaussian averages i.e. there is a constant $\tilde{B}_p$ (resp $\tilde{A}_p$) such that

$$(E \| \sum_{j=1}^{m} x_j r_j \|^p)^{\frac{1}{p}} = \int_{-\infty}^{+\infty} \cdots \int_{-\infty}^{+\infty} (\| \sum_{j=1}^{m} r_j x_j \|^p \, e^{-\frac{1}{2} \sum_{j=1}^{m} s_j^2} / (2\pi)^{\frac{n}{2}}) ds_1 \ldots ds_n$$

$$\geq B_p (\sum_{j=1}^{m} \|x_j\|^2)^{\frac{1}{2}}$$

$$(\text{resp.} \quad (E \| \sum_{j=1}^{m} x_j r_j \|^p)^{\frac{1}{p}} \leq A_p (\sum_{j=1}^{m} \|x_j\|^2)^{\frac{1}{2}}$$

for all $x_1, x_2, \ldots, x_m$ in X $(m = 1, 2, \ldots)$.

References

[1] Ed Dubinsky, A.Pełczyński and H.P.Rosenthal, On Banach spaces X for which $\pi_2(_\infty, X) = B(_\infty, X)$, Studia Math. 44(1972), 617-648.

[2] Y.Gordon and D.Lewis, Absolutely summing maps and local unconditional structures , Acta Math., to appear.

[3] T.Hoffman-Jorgensen, Sums of independent Banach space valued random variables, Aarhus University, Preprint series 1972-1973, No 15, cf. also Studia Math. 52(1974), 159-186.

[4] W.B.Johnson, On finite dimensional subspaces of Banach spaces with local unconditional structure, Studia Math. 51(1974), 225-240

[5] J.P. Kahane, Some random series of functions, Heath Math. Mono. Lexington, Mass. 1968.

[6] S.Kwapień, Isomorphic characterization of inner product spaces by orthogonal series with vector valued coefficients, Studia Math. 44 (1972), 583-595.

[7] S.Kwapień, Isomorphic characterization of inner product spaces..., Seminaire Maurey-Schwartz, Ecole Polytechnique 1972-1973.

[8] J.Lindenstrauss and H.P.Rosenthal, The $\mathcal{L}_p$-spaces, Israel J. Math. 7(1969), 325-349.

[9] B.Maurey, Théorèmes de factorisation pour les opérateurs linéaires à valeurs dans les espaces L^p, Asterisque 11(1974).

[10] G.Nordlander, On sign-independent and almost sign-independent convergence in normed linear spaces, Ark. Mat. r(1962), 287-296.

[11] W.Orlicz, Über unbedingte Konvergenz in Funktionräumen (I), (II), Studia Math. r(1933), 33-37; 41-47.

[12] H.P.Rosenthal, On subspaces of L^p, Annals of Math. 97 (1973), 344-373.

[13] Seminaire Maurey-Schwartz, École Polytechnique, Paris 1972 - 1973.

[14] Nicole Tomczak-Jaegerman, The module of smoothness and convexity and the Rademacher averages of trace classes S_p $(1 \leq p < \infty)$ Studia Math. 50(1974), 163-182.

WINTER SCHOOL
ON PROBABILITY
Karpacz 1975
Springer's LNM

472

TOPICS IN ERGODIC THEORY

By C.Ryll-Nardzewski
Wrocław University

1. Ergodic theorems for subsequences.

The first part of the lecture will be devoted to the ergodic theorems for subsequences. It is based on the ideas contained in the paper by A. Brunel and M. Keane [1].

Fix the probability space (Ω, B, m) in which a measurable, measure preserving (by pre-images) transformation $T : \Omega \to \Omega$ is given.

In the individual ergodic theorem one considers the almost everywhere convergence of the averages

$$\frac{1}{n} \sum_{i=1}^{n} f(T^i w) \quad ,$$

where $f \in L_1(m)$. By contrast the mentioned above authors deal with the averages of the following sort

$$\frac{1}{n} \sum_{i=1}^{n} f(T^{k_i} w) \quad ,$$

where $k_1 < k_2 < \dots$ is an increasing subsequence of the sequence of positive-integer indices. They distinguish the so-called uniform sequences $\{k_i\}$ which are obtained by means of the special construction called by them "apparatus". This construction will be described below.

Let X be a compact space and let the mapping $\varphi : X \to X$ be a homeomorphism.

DEFINITION (X,φ) is said to be a uniformly L - stable (Liapunov system if the family $\{\varphi^n\}_{n\in Z}$ is a uniformly equicontinuous family family of mappings (here Z stands for the set of integers).

The uniform equicontinuity is here to be understood in the sense of the natural uniform structure connected with the compact space. Namely, if U,V are arbitrary open sets containing the diagonal D of the space $X \times X$ i.e. the set

$$D = \left\{(x,x) : x \in X\right\},$$

then the uniform equicontinuity of the family $\{\varphi^n\}$ means that

$$\bigwedge_U \bigvee_V \left\{(x,y) \in V \Rightarrow \bigwedge_{n\in Z} (\varphi^n x, \varphi^n y) \in U\right\} .$$

The notion of the L-stable system was introduced in the years 1951-52 by Oxtoby and Fomin.

Now we add another assumption on the system (X,φ)

Assumption. There exists an orbit dense in X i.e. there exists a point $x_0 \in X$ such that the set $\{\varphi^n x_0\}_{n\in Z}$ is dense in X.

For L - stable systems with a dense orbit the following Oxtoby's [2] theorem holds true

THEOREM 1. The uniformly L - stable system with a dense orbit is strictly ergodic i.e. there exists the unique φ - invariant (and so ergodic) normalized Radon measure μ. Moreover, for any $x \in X$ and any function $f \in C(X)$ we have the formula

$$\int f d\mu = \lim_n \frac{1}{n} \sum_{i=0}^{n-1} f(\varphi^i x) ,$$

i.e. there exists the limit on the right-hand side and the equality holds. Moreover the convergence is uniform in x.

By Radon measure we mean here a regular Borel measure.

We sketch the proof of the theorem. For $f \in C(X)$ we put $Af = f \circ \varphi$. Of course, A is a linear, continuous operator. Now, we utilize the following well known theorem:

if a linear operator A acting in a Banach space has the property: "the set $\{A^j f\}_{j\in Z}$ is conditionally compact for any vector f" then the sequence

$$\left\{ \frac{1}{n} \sum_{j=1}^{n} A^j f \right\}$$

is convergent in norm to an invariant vector $g = Ag$.

This theorem is an easy corollary to the theorem 8.5.1 of [3] and the Banach-Steinhaus theorem on sequences of operators.

By our assumptions the family of functions $\{f \circ \varphi^j\}$ is uniformly bounded and equicontinuous, so that by Arzela's theorem it is conditionally compact. The theorem quoted above assures that the sequence

$$\left\{ \frac{1}{n} \sum_{j=1}^{n} f(\varphi^j x) \right\}$$

satisfies the uniform Cauchy condition so that

$$\bar{f}(x) \overset{df}{=} \lim \frac{1}{n} \sum_{j=1}^{n} f(\varphi^j x)$$

does exist and is a continuous function. Because $\bar{f}$ has the property

$$\bar{f}(x) = \bar{f}(\varphi^j x)$$

for each $j \in Z$ and $x \in X$, from the existenceof a dense orbit in X it follows that $\bar{f} = \text{const}$.

So the correspondence $f \to \bar{f}$ is a linear continuous functional and by the Riesz theorem we get the existence of a regular Borel measure μ such that

$$\bar{f}(x) = \text{const} = \int_X f d\mu .$$

The φ - invariance of the measure μ is implied by the formula $\overline{f}(\varphi x) = \overline{f}(x)$, what in the language of integrals means that

$$\int f d\mu = \int f \circ \varphi \, d\mu$$

for all continuous functions $f \in C(X)$. Were ν another φ - invaria probability measure, the functional

$$L(f) \overset{df}{=} \int f d\nu$$

would also be φ - invariant i.e.

$$L(f) = L(f \circ \varphi) = L(Af)$$

so that from the definition of the function $\overline{f}$ and the continuity of the functional L it would follow that

$$L(f) = L(\overline{f}) .$$

However that means that

$$L(f) = \int f \, d\nu = \int \overline{f} \, d\nu = \overline{f} \int d\nu = \int f \, d\mu$$

and the equality of integrals for all $f \in C(X)$ implies the equality of measures what ends the proof of the theorem.

Now in the space X we distinguish a subset $Y \subset X$ and a fixed element $y_0 \in X$. Consider the orbit of the point y_0 i.e. the set $\{\varphi^j y_0\}_{j \geq 1}$. Let $a(j) = \chi_Y(\varphi^j y_0)$ and let $\{k_i\}$ be the sequence of positive integers exhibiting at which places in the sequence $\{a(j)\}$ there appears "one". So $a(k_i) = 1$, $(i = 1,2,\ldots)$, $a(j) = 0$ $(j \neq k_i)$. Let us make the <u>additional assumption</u> about the set Y: Y is a Jordan set i.e. Y is measurable with respect to the measure $\overline{\mu}$ (the completion of the measure μ appearing in the Oxtoby's theorem), $\overline{\mu}(Y) > 0$, $\mu(\partial Y) = 0$ where ∂Y is the boundary of Y .

DEFINITION. The sequence $\{k_i\}$ of increasing positive integers is called the <u>uniform sequence</u> if there exist an "apparatus" producing it. That means that there exists a strictly ergodic Liapounov system (X,φ) in which one can distinguish a point $y_0 \in X$ and a subset $Y \subset X$ which is a Jordan set with respect to the measure μ obtained from the Oxtoby's theorem that the sequence $\{k_i\}$ determines the appearances of "ones" in the sequence $a(j) = \chi_Y(\varphi^j y_0)$ i.e.

$$a(j) = 1 \iff j \in \{k_i\}$$

The sequence may be finite or not. For classes of sequences $\{a(j)\}$ considered in detail later on we make an additional assumption concerning the frequency of 1's .

<u>Assumption</u>. $\frac{1}{n} \sum_{j=1}^{n} a(j) \to d > 0$.

The uniform sequences do satisfy this assumption because, it is easy to see that

$$\frac{1}{n} \sum_{j=1}^{n} a(j) \underset{n}{\to} \overline{\mu}(Y) .$$

It follows also from the Theorem 6 that will be formulated later on in this lecture.

For other examples of sequences $\{a(j)\}$ the existence of the above mentioned average will be proved later.

Under the above assumption we shall show the validity of

THEOREM 2. <u>For zero-one sequences</u> $\{a(i)\}$

$$d \lim_n \frac{1}{n} \sum_{i=1}^{n} f(T^{k_i} w) = \lim_n \frac{1}{n} \sum_{i=1}^{n} f(T^i w)\, a(i)$$

<u>The existence of any one of the above limits implies the existence of another and their equality</u>.

In this way the problem of Brunel and Keane's averages has been reduced to the problem of classical ergodic averages with "weights" $\{a(j)\}$.

<u>Proof</u>. Let $n = k_s$. Then, if the limit on the right exists then

$$\frac{1}{n} \sum_{j=1}^{n} f(T^j w) a(j) = \frac{1}{n} \sum_{i=1}^{s} f(T^{k_i} w) = \frac{1}{k_s} : s . \frac{1}{s} \sum_{i=1}^{s} f(T^{k_i} w)$$

$$\xrightarrow[s]{} d. \lim_{s} \frac{1}{s} \sum_{i=1}^{s} f(T^{k_i} w) .$$

Conversely, if the limit on the left exists then

$$\frac{1}{n} \sum_{j=1} f(T^j w)a(j) = \frac{1}{n} \sum_{k_i \leq n} f(T^{k_i} w) = \frac{b(n)}{n} \cdot \frac{1}{b(n)} \sum_{i=1}^{b(n)} f(T^{k_i} w)$$

where

$$b(n) = \sum_{j=1}^{n} a(j) .$$

Now, the assertion is obvious because $b(n)/n \xrightarrow[n]{} d$ by our assumption.

Example (of the L-stable system). Let $K = \{z \in C : |z| = 1\}$ be the unit circle, $\varphi(z) \overset{df}{=} ze^{i\xi}$, where ξ is uncommensurable with π. Then the orbit of each point is dense in the space K.

The unit circle is an example of a monothetic group i.e. a topological group which possesses a dense cyclic subgroup.

Any monothetic group G together with an element $u \in G$ for which $\overline{\{u^n\}}_{n \in Z} = G$ is an example of a uniformly L - stable system if one defines $\varphi(x) = xu$. For the circle group the family $\{\varphi^n(t) = t + n\}$ (mod 2π) is the uniformly equicontinuous family of functions with respect to t.

We have the following

THEOREM 3 ([4]). *Any uniformly L - stable system (X,φ) with a dense orbit is isomorphic with the system (G,T) where G is a compact monothetic topological group and T is the shift operator, shifting by an element $u \in G$ the powers there-of form a dense set in G. Under this isomorphism the measure μ, the existence there-of being assured by Oxtoby's theorem, corresponds to the Haar measure on G.*

The plan of the proof is clear because the isomorphism should map the element with the dense orbit into the element u. The rest of the proof consists of the extension of the isomorphism onto the whole space X. That requires some tedious work which we omit here.

However, the above theorem justifies the necessity of the description of monothetic compact groups (cf. e.g. W. Rudin, Fourier analysis on groups, p. 40) showing at the same time that the assumptions concerning the Brunel and Keane's "apparatus" are very strong indeed and the class of uniformly L - stable systems is not very big.

For a monothetic compact group G and the element $u \in G$ generating the dense cyclic subgroup in G the homomorphism

$$\hat{G} \ni \chi \to \chi(u) \in K$$

of the character group $\hat{G}$ into the unit circle is an isomorphism. In fact, if for a character $\chi_0 \in \hat{G}$, $\chi_0(u) = 1$ then $\chi_0(u^n) = 1$ for each n. Because χ_0 is the continuous function and $\{u^n\}_{n\in Z}$ form the set dense in G we have that $\chi_0 = 1$. Thus $\hat{G}$ is isomorphically embedded into K_d the unit circle $K = \{z \in C : |z| = 1\}$ equipped with the discrete topology.

Conversely, if G is a topological group such that $\hat{G} \subset K_d$ then G is the homomorphic image "onto" of the group $(K_d)^\wedge$ by means of the mapping

$$(K_d)^\wedge \ni \chi \overset{h}{\to} \chi/\hat{G} \in (\hat{G})^\wedge = G$$

$(K_d)^\wedge$ is the compact monothetic group. The homomorphic image of a monothetic group is also a monothetic group (if u is an element which generates a dense subgroup then in the image the same rôle is played by the element $h(u)$) so that K_d determines the "largest" monothetic compact group $(K_d)^\wedge$ the homomorphic images thereof give all possible monothetic compact groups. The fact that any monothetic compact group may be obtained in such a way follows from the first part of the reasoning. Denote by $(K_d)^\wedge = Z_{Bohr}$ (the Bohr compactification of the group of integers Z). Z_{Bohr} is described by the following conditions

1° Z_{Bohr} is compact, Z is dense in Z_{Bohr},

2° Each character of the group Z extends to the continuous character of the group Z_{Bohr}.

We shall write down the monothetic group Z_{Bohr} as $\langle Z_{Bohr}, 1\rangle$ in order to underline that the iterations of 1 (in the additive notation!) are dense in the group Z_{Bohr}. All the monothetic compact groups are of the form $\langle h(Z_{Bohr}), h(1)\rangle$ where h is a continuous homomorphism.

If $Y \subset h(Z_{Bohr})$ and if Y is a Jordan set with respect to the Haar measure in $h(Z_{Bohr})$ then $h^{-1}(Y)$ is a Jordan set in Z_{Bohr}.

it follows from the fact that h is continuous and open as a quotient mapping

$$h : (K_d)^{\wedge}/_{\mathrm{Anh}\,\hat{G}} \longleftrightarrow G$$

Such a mapping transforms the Haar measure into the Haar measure and, moreover, preserves the boundary operation, and that is what is needed in the definition of a Jordan set. Eventually, we have

THEOREM 4. _The zero-one sequence_ $\{a(j)\}$ _determining the Brunel and Keane's uniform sequence_ $k_1 < k_2 < \ldots$ _may be described as follows:_

There exists a subset $Y \subset Z_{Bohr}$, Y _being a Jordan set with respect to the Haar measure, for which_ $a(j) = \chi_Y(j)$ _or in other words_ $\chi_Y(j) = \chi_Y(0+j.1)$.

The 0 of the group Z_{Bohr} is here the starting point. The property "to be a Jordan set" is preserved under shifts because we utilize the Haar measure which is invariant under translations. Hence we may start from an arbitrary point.

As examples of sequences $\{a(j)\}$ for which the limits of ergodic averages with "weights" $\{a(j)\}$ do exist (what is equivalent to the existence of limits of Brunel and Keane's averages) we may give the "bounded Besicovitsch sequences" i.e. the sequences of complex numbers $\{a(j)\}_{j\in N}$ with the properties:

1^0 $|a(j)| \leq \alpha$, α - a constant;

2^0 for any $\varepsilon > 0$ there exists a trigonometric polynomial

$$w_\varepsilon(\cdot) = \sum_s \gamma_s \, e^{i\vartheta_s j}$$

such that

$$\overline{\lim_n} \; \frac{1}{n} \sum_{j=1}^{n} |a(j) - w_\varepsilon(j)| \leq \varepsilon \quad .$$

Here $j \in N$, $0 \leq \vartheta_s < 2\pi$, $\sum_s$ is a finite sum. Now, we prove two theorems giving applications of "bounded Besicovitsch sequences".

THEOREM 5. <u>For any function</u> $f \in L_\infty(m)$ <u>and any bounded Besicovitsch sequence</u> $\{a(j)\}$

$$\lim_n \frac{1}{n} \sum_{j=1}^{n} f(T^j w)\, a(j) = \bar{f}(w)$$

<u>exists</u> m - <u>almost everywhere and</u> $\bar{f} \in L_1(m)$.

<u>Proof</u>. Put $a(j) = e^{i\vartheta j}$, $0 \leqslant \vartheta < 2\pi$ being fixed. We shall show that the averages

$$\frac{1}{n} \sum_{j=1}^{n} f(T^j w)\, e^{i\vartheta j}$$

are convergent for m - almost all w .

Indeed, in the space $(\Omega m, T) \times (K, \lambda, \vartheta)$, where K is the unit circle, λ - the normalized Lebesgue measure and ϑ - the rotation by the angle ϑ , we define the function

$$g(w,z) = f(w)z \quad .$$

By virtue of the individual ergodic theorem applied to the tranformation $T' : (w,z) \to (Tw,\ ze^{i\vartheta})$ the averages

$$\frac{1}{n} \sum_{i=1} f(T^j w) e^{i\vartheta j}\, z$$

are convergent for almost all points (w,z) . Thus, using the Fubini theorem, one takes such a $z \in K$ for which the convergence of the above averages takes place for almost all w (and λ - almost all z have this property). Because z is here the constant parameter, we get the m - almost everywhere convergence of the averages

$$\frac{1}{n} \sum_{j=1}^{n} f(T^j w) e^{i\vartheta j} \ .$$

So we immediatelly get our assertion in the case when $a(j) = w(j)$, ($w(.)$ being a trigonometric polynomial) because the operation of taking the limit averages is a linear operator.

Finally, if $f \in L_\infty(m)$ and $\{a(j)\}$ is a bounded Besicovitsch seq.

$$\overline{\lim_n} \frac{1}{n} \sum_{j=1}^{n} |f(T^j w)| \, |a(j) - w_\varepsilon(j)| \leq \overline{\lim_n} \frac{1}{n} \sum_{j=1}^{n} \|f\|_\infty |a(j) - w_\varepsilon(j)| \leq \varepsilon \|f\|_\infty ,$$

that, because of arbitrariness of ε and the existence of the average for "weights" $\{w_\varepsilon(j)\}$ give the existence of the desired average for "weight" $\{a(j)\}$

THEOREM 6. <u>The assertion of the Theorem 5 remains valid for any function</u> $f \in L_1(m)$.

<u>Proof</u>. The assertion has been already proven for those $f \in L_1(m)$ which are bounded. To prove it for any integrable function f we shall make use of the following Banach - Mazur theorem on sequences of linear mappings: If $A_n : E \to S(\Omega, m)$ $(n = 1,2,\dots)$ is a sequence of type (F) (linear, metric, complete, thus in particular the space $L_1(m)$) into the space of functions measurable with respect to the measure m and possesing the following properties

a) in the space E there exists a dense subset E_0 such that for any $x \in E_0$ the sequence $\{A_n(x)(w)\}$ is convergent for m - almost all w ;

b) for each $x \in E$, $\sup_n |A_n(x)(w)| < \infty$ for m - almost all w, then for each $x \in E$ the sequence $\{A_n(x)(w)\}$ converges m - almost everywhere.

Define

$$(A_n f)(w) = \frac{1}{n} \sum_{j=1}^{n} f(T^j w) a(j)$$

The condition a) is satisfied because of the Theorem 1 and the condition b) is implied by the evalution

$$\left|\frac{1}{n} \sum_{j=1}^{n} f(T^j w) a(j)\right| \leq \alpha \frac{1}{n} \sum_{j=1}^{n} |f(T^j(w)|$$

and the individual ergodic theorem applied to the function $|f|$ .This ends the proof.

While considering the averages of the type

$$\frac{1}{n}\sum_{j=1}^{n} f(T^j w)a(j)$$

we wanted to find a possibly large, class of sequences $\{a(j)\}$ for which the averages are convergent almost everywhere. In this respect, the two above theorems distinguish"bounded Besicovitsch sequences" from which we choose the following classes of examples:

(1) trigonometric polynomials $w(j)$,

(2) their uniform limits,

(3) almost periodic Bohr sequences,

(4) Fourier coefficients of complex measures on the circle i.e

$$a(j) = \hat{\lambda}(j) = \int_0^{2\pi} e^{-i\vartheta j}\, \lambda(d\vartheta) \quad ,$$

where λ is a complex measure on the interval $[0, 2\pi]$.

It is evident that the sequences belonging to the first three classes may serve as the weights in the Brunel and Keane's averages. The fourth class deserves the special attention.

The complex measure λ may be decomposed into the continuous (i.e. atomless) part λ_c and the discrete part λ_d , $\lambda = \lambda_c + \lambda_d$.

$$\lambda_d = \sum_k \alpha_k \delta_{\vartheta_k} \quad , \qquad\qquad \sum |\alpha_k| < \infty \ .$$

Therefore

$$\hat{\lambda}_d(j) = \sum_{k=1}^{\infty} \alpha_k e^{i\vartheta_k j} \quad ,$$

and that is the almost periodic Bohr sequence.

Dealing with the continuous part λ_c we are going to use the classical fact that

$$\frac{1}{n}\sum_{j=1}^{n} |\hat{\lambda}_c(j)|^2 \underset{n}{\rightarrow} 0 \ .$$

By the Schwartz inequality for sequences we get that

$$\frac{1}{n}\sum_{j=1}^{n}|\hat{\lambda}_c(j)| \underset{n}{\to} 0 \quad ,$$

what proves that $\{\hat{\lambda}_c(j)\}$ is a bounded Besicovitsch sequence (for each $\varepsilon > 0$ one should put $w_\varepsilon(.) = 0$).

The formula we utilized above may be proved as follows:

If η is a finite, complex Borel measure on the interval $[0,2\pi)$ then for its Fourier coefficients

$$\hat{\eta}(j) = \int_{[0,2\pi)} e^{i\vartheta j}\, d\eta$$

one has

$$\frac{1}{n}\sum_{j=1}^{n}\hat{\eta}(j) \to \eta(\{0\}) \quad ,$$

what is implied by the fact that

$$\frac{1}{n}\sum_{j=1}^{n} e^{i\vartheta j} \to \begin{cases} 0 & 0 < \vartheta < 2\pi \\ 1 & \vartheta = 0 \quad . \end{cases}$$

However, $|\hat{\lambda}_c(j)|^2 = (\lambda * \lambda^*)\hat{}(j)$ where λ^* is the measure defined by the formula $\lambda^*(A) = \overline{\lambda(-A)}$.

Therefore, for the measure $\eta = \lambda * \lambda^*$, it follows from the above remarks that

$$\frac{1}{n}\sum_{j=1}^{n}|\hat{\lambda}_c(j)|^2 = \frac{1}{n}\sum_{j=1}^{n}(\lambda_c * \lambda_c^*)\hat{}(j) \underset{n}{\to} \lambda_c * \lambda_c^*(\{0\}) \quad .$$

Now, one should use the fact that for any measure η

$$(\eta * \eta^*)\ (\{0\}) = \sum_k |\alpha_k|^2 \quad ,$$

where α_k is the (complex) mass of the k-th atom of measure m . Because the measure λ_c , being the continuous measure, has no atoms we get

$$(\lambda_c * \lambda_c^*)(\{0\}) = 0 \quad ,$$

what implies the desired condition.

For Besicovitsch sequences the average

$$\underset{n}{M}(a(n)) \overset{df}{=} \lim_n \frac{1}{n} \sum_{j=1}^{n} a(j) \quad ,$$

exists because it exists for trigonometric polynomials (that comes out of an easy computation) and moreover the condition 2^0 from the definition of Besicovitsch sequences is satisfied.

The sequence

$$\gamma(\vartheta) \overset{df}{=} \underset{n}{M} (a(n) e^{-i\vartheta n})$$

is called the sequence of Fourier coefficients of the sequence $a(.)$. The class of bounded Besicovitsch sequences together with the function of two variables

$$(a(.), b(.)) = \underset{n}{M} [a(n) \overline{b(n)}]$$

forms a pre-Hilbert space so that we have here the convenience of the following Bessel inequality:

for any, but different from each other, $\vartheta_1, \vartheta_2, \ldots, \vartheta_k$

$$\sum_{s=1}^{n} |\gamma(\vartheta_s)|^2 \leq \underset{n}{M} (|a(n)|^2) < \infty \quad .$$

Directly from this inequality it follows that $\gamma(\vartheta) \neq 0$ only for countably many values of ϑ.

So, to each Besicovitsch sequence we can attach the formal Fourier series

$$a(.) \approx \sum_s \gamma(\vartheta_s) e^{i\vartheta s} \quad ,$$

for which (from the Bessel inequality)

$$\sum_s |\gamma(\vartheta_s)|^2 < \infty \ .$$

The functions $\chi(.) = e^{i\vartheta_s \cdot}$ are, at the same time, the characters of the groups Z and Z_{Bohr} (by virtue of the definition of the latter) and form in Z_{Bohr} the orthonormal basis. Utilizing the sequences $\gamma(\vartheta_s)$ appearing in the above formulas one can define the function

$$f(.) = \sum_s \gamma(\vartheta_s)\ e^{i\vartheta_s \cdot} \ ,$$

the series being convergent in $L_2(Z_{Bohr})$ .The series is convergent because by the Riesz-Fisher theorem and the condition $\sum_s |\gamma(\vartheta_s)|^2 < \infty$ the function $f \in L_2(Z_{Bohr})$.

One can even prove that $f \in L_\infty(Z_{Bohr})$, $L_\infty(Z_{Bohr})$ being equipped with the Haar measure. Besicovitsch proved the theorem saying that any $f \in L_\infty(Z_{Bohr})$ may be obtained in the above mentioned way i.e. by extention of the properly chosen bounded Besicovitsch sequence. Unfortunaly, what here makes trouble is that the correspondence $a(.) \to f$ is not one-to-one.

If the function f is an extension of a Besicovitsch sequence $\{a(j)\}$, and the function g that of $\{b(j)\}$, then for the sequence $\{a(j)\cdot b(j)\}$ (which is also a Besicovitsch sequence) the function $f(.)g(.)$ gives the necessary extension. That is relatively obvious if $a(j) = w_1(j)$, $b(j) = w_2(j)$ where w_1, w_2 are trigonometric polynomials because then $\gamma_3(\vartheta)$ i.e. the coefficient of $e^{in\vartheta}$ in the function $w_1 \cdot w_2$ is the convolution of the coefficients $\gamma_1(\vartheta)$ of w_1 and $\gamma_2(\vartheta)$ of w_2 . However, that means that respective series are multiplied. Of course for arbitrary Besicovitsch sequences the above remark requires a seperate proof which we skip in this lecture.

We shall use the remark in the case of zero-one Besicovitsch sequences. If $a(j) = 0$ or 1 then the condition $a^2(j) = a(j)$ is satisfied. Therefore the function f obtained from the sequence $\{a(.)\}$ satisfies condition $f^2(.) = f(.)$ and thus takes only 0 or 1 as its values. Thus the function f is an indicator $f(.) = \chi_Y(.)$, where $Y \subset Z_{Bohr}$, Y is the measurable set with respect to μ and μ is the Haar measure on Z_{Bohr} . The uniform sequences of Brunel and Keane required for their definition the sets that are measurable in the sense of Jordan. For Jordan sets the following lemma is valid:

<u>If the set</u> S <u>is dense in the Jordan sets</u> Y_1 <u>and</u> Y_2 <u>and</u>

$Y_1 \cap S = Y_2 \cap S$ <u>then</u> $\bar{\mu}(Y_1 \triangle Y_2) = 0$.

For the Lebesque measure this situation, of course, can not happen and that is why the correspondence between the set $Y \subset Z_{Bohr}$ and the Besicovitsch sequence is made by means of Fourier coefficients and not by means of indicators.

The following question is related to the problem of weighted averages:

What should be demanded of the transformation T in order that, for any bounded Besicovitsch sequence $\{a(j)\}$ and any integrable function $f \in L_1(m)$, the limit of weighted averages be constant

$$\lim_n \frac{1}{n} \sum_{j=1}^{n} f(T^j w)\, a(j) = \text{const.}$$

Here we have

THEOREM 7. <u>The ergodic averages weighted by means of bounded Besicovitsch sequences are constant everywhere if, and only if the transformation</u> T <u>is a weak mixing i.e. if for any Borel sets</u> A,B

$$\frac{1}{n} \sum_{j=1}^{n} |m[T^{-j}(A) \cap B] - m(A)\, m(B)| \underset{n}{\to} 0 \quad .$$

The property of weak mixing may, equivalently, be expressed by two conditions ([6]):

a) T is an ergodic transformation,

b) the operator T defined by the formula $(Tf)(w) \overset{df}{=} f(Tw)$ has no eigenvalues $\lambda \neq 1$ i.e. the equality $f(Tw) = \lambda f(w)$ holds only for the function $f \equiv 0$.

In the proof of the above theorem we shall make use of the second definition of the weak mixing.

<u>Proof</u>. First, we establish the necessity of the condition. Put $a(j) \equiv 1$. That is, of course, a bounded Besicovitsch sequence and for this sequence, by our assumption

$$\frac{1}{n} \sum_{j=1}^{n} f(T^j w) \underset{n}{\to} \text{const}$$

for any integrable function $f \in L_1(m)$, and that means that the transformation T is ergodic ([6]). Next put

$$a(j) = e^{i\vartheta j} = \lambda^j , \qquad 0 < \vartheta < 2\pi , \quad \lambda = e^{i\vartheta} .$$

Then by our assumption

$$\frac{1}{n} \sum_{j=1}^{n} f(T^j w) \lambda^j \underset{n}{\to} \bar{f}(w) \equiv \text{const}$$

m - almost everywhere.

Writing down the average on the left differently we get the formula $\bar{f}(Tw)\lambda = \bar{f}(w)$ and because $\lambda \neq 1$ and (by assumption) $\bar{f}(w) = \text{const}$, it follows that $\bar{f} \equiv 0$. However, from the formula $\bar{f}(Tw)\lambda = \bar{f}(w)$ written in the form $T(\bar{f})(w) = \lambda^{-1}\bar{f}(w)$ it follows that $\bar{f}$ is the eigenfunction of the operator T corresponding to the eigenvalue λ^{-1} . Because $\bar{f} \equiv 0$ we eventually get that T is a weak mixing.

We start the proof of sufficiency in the case $a(j) \equiv 1$. Then, from the ergodicity of the transformation T it follows that the averages tend to a constant. If $a(j) = e^{i\vartheta j}$, $0 < \vartheta < 2\pi$, the second condition from the definition of the weak mixing intervenes and the weighted averages, as was shown above, tend to zero.

If $a(j) = w(j)$, where $w(.)$ is a trigonometric polynomial then the averages weighted by means of the sequence $\{w(j)\}$ tend to a constant.

An arbitrary bounded Besicovitsch sequence may be approximated by trigonometric polynomials so that the average is also being approximated by constants and therefore is constant itself.

Now, we quote two results that do not deal directly with uniform sequences but, nevertheless, are closely related to the problems dealt with by Brunel and Keane.

THEOREM 8. (Blum and Hanson [7], [1]). Let $k_1 < k_2 < \dots$ be an arbitrary sequence of positive integers. Then for any function

$$f \in L_p(m) , \quad 1 \leq p < \infty$$

$$\lim_n \frac{1}{n} \sum_{j=1}^{n} f(T^{k_j} w) = \int f dm$$

__in the norm of__ $L_p(m)$ __whenever__ T __is the strong mixing i.e. whenever for any measurable sets__ $A, B \subset \Omega$

$$m(T^{-j}(A) \cap B) \underset{j}{\to} m(A)\, m(B)$$

__Proof.__ It is enough to restrict attention to the case $f \in L_2(m)$ and to consider only these functions f for which $\int f dm = 0$. Otherwise, we can approximate an arbitrary function from $L_p(m)$ by functions from L_2, and / or we shift the function f by a constant so that the average will be shifted by the same constant.

Denote by $(.|.)$ the scalar product in $L_2(m)$. We have to show that

$$\frac{1}{n} \sum_{j=1}^{n} f(T^{k_j} w) \underset{n}{\to} 0$$

Compute

$$\| \frac{1}{n} \sum_{j=1}^{n} f(T^{k_j} w)\|_2^2 = \frac{1}{n} \sum_{p=1}^{n} \frac{1}{n} \sum_{q=1}^{n} (f|T^{k_p-k_q} f) .$$

The above way of writing $T^{k_p-k_q} f$ tacitly assumes that T is an invertible transformation. However, in the case when T is not an automorphism of the space Ω the summands

$$(f|T^{k_p-k_q} f)$$

for $k_p-k_q < 0$ ought to be replaced by

$$(T^{k_q-k_p} f|f) .$$

Now, we use an alternative definition of the strong mixing ([6]). The transformation T is the strong mixing if for any functions $g,h \in L_2(m)$ we have

$$(g|T^n h) \underset{n}{\rightarrow} (g|1)(1|h) \ .$$

Let us put $g = h = f$. Then $(f|1) = \int f dm = 0$, so that denoting $a_n = (f|T^n f)$, by virtue of the properties of the transformation T, we get that $a_n \to 0$ and $a_n \to 0$ as $n \to \infty$.

The rest of the proof is contained in the following Lemma which is of purely algebraic character:

LEMMA. *If*

$$\lim_{|n| \to \infty} a_n = 0$$

and $k_1 < k_2 < \dots$ *is an arbitrary sequence of positive integers then*

$$\frac{1}{n^2} \sum_{p,q=1}^{n} (a_{k_p - k_q}) \underset{n}{\rightarrow} 0 \ .$$

We put in the formula above $(a_n) = \bar{a}_{-n}$ for $n < 0$.

The second promised result is related to the similar problems in Hilbert spaces. Namely, let E be a Hilbert space, T a unitary operator acting in E, $f \in E$. We are looking for the class of such sequences $\{a(j)\}$ for which the limit in norm of the following averages exists:

$$\frac{1}{n} \sum_{j=1}^{n} T^j f \, a(j)$$

Here we have the following

THEOREM 9. *The norm limit of the above averages exists whenever the sequence* $\{a(j)\}$ *satisfies the following conditions:*

(i) *the sequence*

$$S_n(\vartheta) = \frac{1}{n} \sum_{j=1}^{n} a(j) \, e^{i\vartheta j}$$

<u>is bounded by a constant independent of</u> n <u>and of</u> ϑ ;

(ii) $\{S_n(\vartheta)\}$ <u>is pointwise convergent for each</u> ϑ .

<u>Proof</u>. The proof is based on the spectral theorem. Namely, let $M(d\vartheta)f$ denote the "differential" of the spectral measure of the unitary operator T on the fixed vector $f \in E$ i.e. of the vector measure with orthogonal values (on disjoint Borel subsets of the interval $[0,2\pi)$).

We check that the sequence of averages in question satisfies Cauchy condition. Indeed, we have

$$\left\|\frac{1}{n}\sum_{j=1}^{n} T^j a(j)f - \frac{1}{m}\sum_{j=1}^{n} T^j a(j)f\right\|^2 = \int_{[0,2\pi)} |S_n(\vartheta) - S_m(\vartheta)|^2 (M(d\vartheta)f,f)$$

where $(M(d\vartheta)f,f)$ is the non-negative, finite Borel measure on the interval $[0,2\pi)$. By virtue of assumptions (i) and (ii) about the sequence $\{S_n(\vartheta)\}$, and by the Lebesgue theorem, we get that the Cauchy condition is fulfilled. This ends the proof.

The above theorem remains true also in the case if T is a contraction i.e. $\|T\| \leq 1$. This case may be reduced to the previous one by means of the Nagy's theorem which says that for any contraction operator T acting in the Hilbert space E there exists a unitary dilation U acting in the Hilbert space H containing E such that

$$T^n f = PU^n f$$

for each vector $f \in E$, where P denotes the orthogonal projector of the space H onto the subspace E .

Perhaps now, in the conclusion, we owe to the reader an explanation why no theorem, from the quoted above paper by Brunel and Keane [1], and dealing with the convergence of averages weighted by uniform sequences, appeared above. Brunel and Keane proved the theorem analogous to the Theorem 6, the words "bounded Besicovitsch sequence" being replaced in their paper by the words "uniform sequence". The following shows however that the Theorem 6 is stronger.

LEMMA. <u>Every uniform sequence is a bounded zero-one Besicovitsch sequence</u>.

We give a sketch of the proof.

Proof. Let $\{a(j)\}$ be a uniform sequence determined by the Jordan set Y and point y_0, $a(j) = \chi_Y(\varphi^j y_0)$,

In view of the regularity of measure μ obtained from the Oxtoby's theorem and from the obvious inequalities

$$\chi_{\mathrm{Int}\,Y} \leq \chi_Y \leq \chi_{\overline{Y}}$$

it follows that if Y is a Jordan set relative to μ (i.e. $\mu(\partial Y) = 0$) then for each $\varepsilon > 0$ there exist continuous functions g and h such that

$$g \leq \chi_{\mathrm{Int}\,Y} \leq \chi_Y \leq \chi_{\overline{Y}} \leq h$$

and

$$\int_X (h - g)d\mu < \varepsilon \, .$$

But the sequences $A(j) = g(\varphi^j y_0)$ and $B(j) = h(\varphi^j y_0)$ as almost periodic sequences are also bounded Besicovitsch sequences.

Because in the Theorem 1 $\int f d\mu$ does not depend on x appearing on the right-hand side of the equality, we put in the Theorem 1 $x = y_0$ and $f = g$, $f = h$ thus getting

$$\int_X g d\mu = \underset{j}{M}\left\{A(j)\right\} \leq \varliminf_n \frac{1}{n} \sum_{j=1}^{n} a(j) \leq \varlimsup_n \frac{1}{n} \sum_{j=1}^{n} a(j) \leq \underset{j}{M}\left\{B(j)\right\} =$$

$$= \int h d\mu \, .$$

Because the extreme terms of the above inequalities may differ arbitrarily little, we get the existence of the average

$$\underset{j}{M}\left\{a(j)\right\} = \lim_n \frac{1}{n} \sum_{j=1}^{n} a(j)$$

In the same way we can show that $\underset{j}{M}\left\{|a(j) - A(j)|\right\} < \varepsilon$, ($\varepsilon$ arbitrary positive) hence $\{a(j)\}$ is a (bounded). Besicovitsch sequence.

2. Right topological groups and distal flows.

The second part of the lecture will be devoted to the research done by R.Ellis, H.Furstenberg and I.Namioka originated back in 1957, and is mainly based on the mimeographed notes by I.Namioka ([8]).

At the beginning we introduce a few notions related to flows.

DEFINITION. The <u>flow</u> is a pair (X,S), where X is called the phase space, and which in our case will be always a compact topological Hausdorff space, and S is an abstract semigroup acting continuously on X . That means that a mapping $(s,x) \to s.x \in X$ is given which satisfies the conditions

(i) for fixed s the mapping is a continuous function of the variable x ,

(ii) $s_1(s_2x) = (s_1s_2)x$.

If (X,S) and (Y,S) are flows on which the same semigroup S acts and if there exists a continuous mapping $f : X \xrightarrow{\text{onto}} Y$ which commutes with the semigroup operation, i.e. $f(sx) = sf(x)$ for arbitrary $s \in S$ and $x \in X$, the flow (Y,S) is called the <u>subflow</u> of (X,S) or, equivalently, we say that the flow (X,S) is an extension of the flow (Y,S) . In the case f is a homeomorphism, the flows (X,S) and (Y,S) are said to be <u>isomorphic</u>.

The notion of the distal flow describes the special way of action of the semigroup S on X .

DEFINITION. The flow (X,S) is said to be <u>distal</u> if for any $x,y \in X$, $x \neq y$,

$$\overline{\{(sx,sy) \in X \times X : s \in S\}} \cap \Delta = \emptyset \ .$$

Here the closure of the orbit of the element (x,y) is understood in the sense of the product topology in $X \times X$, and Δ stands for the diagonal in $X \times X$, i.e. $\Delta = \{(x,x) : x \in X\}$.

Distal flows may alternatively be described as follows: the flow (X,S) is distal whenever the conditions

$$s_\alpha x \to u \ , \qquad s_\alpha y \to u$$

imply that $x = y$. Here α runs over some directed set of indices. Were X the compact metric space with the metric ρ , the fact that the flow is distal would mean that for $x \neq y$

$$\inf_{s \in S} \rho(sx, sy) > 0 .$$

The subset A of the phase space X is said to be s - <u>invariant</u> if for each $s \in S$ and each $x \in A$ $sx \in A$. The flow (X,S) is said to be <u>minimal</u> if the space X does not contain proper closed s - invariant subsets. That means that for each $x \in X$ the orbit $\{sx : s \in S\}$ of the point x is dense in the space X.

The definitions given so far were connected with the notion of the flow. The second bunch of definitions concerns so-called right topological groups. The theorems that will be explained below show the interrelations between those two notions.

DEFINITION. G is said to be a <u>right topological group</u> if it is a group and a topological space (but not necessarily a topological group) such that the mapping

$$G \ni x \to xy \in G$$

is continuous for each $y \in G$.

The right topological group is called <u>admissible</u> if there exists a subset $S \subset G$, dense in G such that the mappings $x \to zx$ are continuous as functions of x for each $z \in S$. If the mappings $x \to s_1x$ nad $x \to s_2x$ are continuous then also the mapping $x \to s_1s_2x$ is continuous. However that means that the subset S may be enlarged to the semigroup generated by S. From now on we shall always assume that S is a sub-semigroup of G. In the right topological groups Namioka ([8]) has introduced so-called σ - topology:

Let G be a right topological group and let the mapping $\varphi : G \times G \to G$ be defined by means of the formula

$$\varphi(x,y) = x^{-1} y$$

The σ - <u>topology</u> is determined by the mapping φ as the quotient topology in G. That means that the set $U \subset G$ is σ - open if, and only if $\varphi^{-1}(U)$ is an open set in $G \times G$, equipped with the product topology.

Let K be a subgroup of the right topological group G. The

homogeneous space $G/K = \{gK : g \in G\}$ will be here equipped with the quotient topology obtained from the original topology in G by means of the canonical mapping $G \to G/K$. The meaning of the σ - topology is explained by the following

THEOREM 1. <u>The subgroup K is σ - closed in the right topological group G if and only if the homogeneous space G/K equipped with the quotient topology is a Hausdorff space</u>.

The proof of this theorem may be found in the notes [8].

The aim of this lecture is to show the interrelations between the distal flows and the right admissible topological groups. The following theorem is due to Ellis [9] and its, perhaps enigmatic, formulation will become clear in the proof, as it requires the introduction of a few new notions

THEOREM 2. <u>Between the distal minimal flows and right topological groups there exists a correspondence of the isomorphism type</u>.

<u>Proof</u>. Let (X,S) be a flow. Let us consider the space X^X of all functions (not necessarily continuous) on the set X into itself, equipped with the Tichonow topology of pointwise convergence. The space X^X may be regarded as a semigroup of mappings (with respect to superposition of transformations). There exists the natural homomorphism $h : S \to X^X$ given by the formula $h(s)x = s \cdot x$. Because (X,S) is a flow, right from the definition we get that the semigroup $h(S) \subset X^X$ consists of the continuous mappings mapping X into X. It is easy to prove that the closure of h(S) in the product topology is also a semigruop. In fact, if $f,g \in \overline{h(S)}$ then for each $x \in X$

$$f(x) = \lim_{\alpha} S_{\alpha} (x)$$

$$g(x) = \lim_{\beta} S_{\beta} (x) \quad ,$$

S_{α}, $S_{\beta} \in h(S)$. But then $S_{\alpha} \cdot S_{\beta} \in h(S)$ and from the continuity of S_{α} it follows that $\lim_{\beta} S_{\alpha}(S_{\beta} x) = s_{\alpha}\, g(x)$ wherefrom $S_{\alpha} g \in \overline{h(s)}$. Taking $\lim S_{\alpha} g(x) = f(g(x)) = (fg)(x)$ we get that $f \cdot g \in \overline{h(S)}$ what was to be proved. R.Ellis has proven ([9]) that the flow (X,S) is distal if and only if $\overline{h(S)}$ is a group.

$G = \overline{h(S)}$ is called the Ellis group of the distal flow (X,S) .

Because $G \subset X^X$, X^X is a compact Hausdorff space, then G as a closed subset of X^X is also a compact Hausdorff space. The homomorphism described above maps the semigroup S into G , $h : S \to G$.

It follows from the properties of topology in X^X that the mapping $g \to gk : G \to G$ the mapping $g \to sg : G \to G$ is continuous for each $s \in h(S)$. Thus G is an admissible compact right topological Hausdorff group, because $h(S)$ is a dense subset in $G = \overline{h(S)}$.

Assume furthermore that the flow (X,S) is minimal. Let x_0 be an arbitrary but fixed element of X . We define the mapping $p : G \to X$ given by the formula $p(g) = gx_0$. From the minimality of the flow it follows that the set $\{sx_0 : s \in S\}$ is dense in X . But $sx_0 = p(h(s))$ and that means that the set $p(h(S))$ is dense in X . Because p is evidently a continuous mapping and $h(S) = G$ we get eventually that $p(G) = X$ i.e. that p is a mapping "onto" .

Let $K = \{g \in G : gx_0 = x_0\}$. K is a subgroup of G . Denote by π the canonical map $\pi : G \to G/K$. Then

$$p(g_1) = p(g_2) \iff \pi(g_1) = \pi(g_2)$$

because

$$p(g_1) = p(g_2) \iff g_1x_0 = g_2x_0 \iff x_0 = g_1^{-1}g_2x_0 \iff g_1^{-1}g_2 \in K$$

$$\iff g_2 \in g_1K \iff \pi(g_1) = \pi(g_2) \quad .$$

The above relation implies the existence of the unique homeomorphism $f : G/K \to X$ such that $p = f \circ \pi$. Because G/K , as the homeomorphic image of the Hausdorff space, is also the Hausdorff space, the Theorem 1 implies that K is a closed subgroup of G in the σ - topology. Moreover, the mapping

$$(s,gK) \to h(s)\, gK : S \times G/K \to G/K$$

defines the flow $(G/K,S)$ isomorphic, by the mapping f , with the flow (X,S) . This isomorphisms makes the correspondence from the formulation of Theorem 2 precise.

Conversely, let G be an admissible, compact right topological Hausdorff group and $S \subset G$ a dense sub-semigroup such that the mappings $x \to zx : G \to G$ are continuous for $z \in S$. Let moreover, K be a σ - closed subgroup of G . Then the semigroup S acts on the set G/K by the mapping $(s,gK) \to sgK$, thus defining the minimal distal flow $(G/K,S)$. It follows from our assumptions that sg is a continuous function of the variable g so that sgK is a continuous function of the variable gK . The minimality of the flow is obvious because for each $g_0 \in G$

$$\left\{\overline{sg_0K : s \in S}\right\} = \left\{ gg_0K : g \in G \right\} = G/K .$$

and the distalness of the flow follows from the following simple computation. If

$$s_\alpha g_1K \underset{\alpha}{\to} gK \quad \text{and} \quad s_2g_2K \underset{\alpha}{\to} gK ,$$

then taking a cluster point of the sequence $\{s_\alpha\}$ (s_0 exists because X is compact) we get that

$$s_0g_1K = s_0g_2K$$

so that

$$g_1K = g_2K$$

References

[1] A.Brunnel and M.Keane, Ergodic theorems for operator sequences, Z.Wahrscheinlichkeitstheorie verw. Geb. 12 (1969), 231-240.

[2] J.C.Oxtoby, Ergodic sets, Bull. Amer. math. Soc. 58 (1952), 116-136.

[3] N.Dunford and J.T.Schwartz, Linear operators, Part I, Wiley -Interscience, New York 1957.

[4] P.R.Halmos and J. von Neumann, Operator methods in classical mechanics II, Ann. of Math. 43 (1942), 332-350.

[5] N.Wiener, Generalized harmonic analysis, Acta Math.55(1930).

[6] P.R.Halmos, Lectures on the ergodic theory

[7] J.R.Blum and D.L.Hanson, On the mean ergodic theorem for subsequences, Bull. Amer. math. Soc. 66 (1960), 308-311.

[8] I.Namioka, Right topological groups, distal flows and a fixed point theorem, 1971, to appear.

[9] R.Ellis, Distal transformation groups, Pacific J. Math. 8 (1958), 401-405.

WINTER SCHOOL
ON PROBABILITY
Karpacz 1975
Springer's LNM

472

A MORE DETERMINISTIC VERSION OF HARRIS - SPITZER'S "RANDOM CONSTANT VELOCITY" MODEL FOR INFINITE SYSTEMS OF PARTICLES

By W. Szatzschneider
Gdańsk University

Introduction

We begin with a summary of results and discussion of the contents of this paper. We shall consider a system of particles interpreted as point masses on the real line. This model will be the one with "random constant velocity" i.e. in which the position of k-th particle is described by the formula $x_k(t) = x_k + v_k \cdot t$, where x_k and v_k are random variables. This system of particles will not be in the macroscopic equilibrium (for discussion of the equilibrium state cf. Ciesielski [3]). We observe a single particle's limit motion with suitable normalization and this leads to an unexpected phenomenon we describe below.

Results

Now, we give the detailed description of our model which we call the model (D). Let N_k be the distribution of random variables $x_k - k$ and U_k the distribution of random variables v_k, $k=0,\pm 1,\pm 2,\dots$ which satisfy the following assumptions:

(D1) $N_k = N$ for $k = \pm 1, \pm 2, \dots$, and N_0 is concentrated at the origin ;

(D2) N_k is symmetric, concentrated on the interval $(-1/2,1/2)$ so that, in particular, $E(x_k - k) = 0$;

(D3) $U_k = U$ for $k = 0, \pm 1, \pm 2, \ldots$;

(D4) U is symmetric, $0 < E|v_k| < \infty$, so that $E(v_k) = 0$;

(D5) $x_0, v_0, x_1, v_1, x_{-1}, v_{-1}, x_2, \ldots$ are independent random variables.

In our model the probability space of the initial conditions is the following:

$$(\Omega,\underline{F},P) = \prod_{-\infty}^{\infty} (R_2,\underline{B},\gamma_k)$$

where $\underline{B}$ is the Borel field in the plane R_2 and $\gamma_k = N_k' \times U_k$, where N_k' is the distribution of x_k .

Let $x_k(t,\omega) = x_k(0,\omega) + v_k(\omega)\cdot t$ for $t \geq 0$, $k = 0,\pm 1,\pm 2,\ldots$, where $x_k(0) = x_k$, and let $y_k(t,\omega)$ for $t \geq 0$ be the actual motion of k-th elastic particle. For fixed (almost every) ω the y_k's may be defined by a deterministic theorem of Harris [6] ($y_k(t,\omega)$ is a polygonal sample path). We restrict our attention to the trajectory $y(t) = y_0(t)$.

THEOREM 1. $Y_A(t) \Longrightarrow x(t)$ <u>as</u> $A \to \infty$, <u>where</u> " $\Longrightarrow$ " <u>means the weak convergence in the space of measures on the space of continuous functions on</u> $[0,1]$, <u>and</u> $Y_A(t) \overset{df}{=} y(At)/\sqrt{A}$. <u>Moreover,</u> $x(t)$ <u>is a Gaussian process determined by conditions:</u>

$$E\,[x(t)] = 0 ,$$

$$E[x(t)x(s)] = \min\,(t,s)E|v| - \frac{1}{2}\,E \min\,[t|u|,\, s|v|] ,$$

<u>where</u> u,v <u>are independent random variables with distribution</u> U .

The proof consists of two parts. The first one is based on the central limit theorem for independent random vectors and the second uses simply the tightness argument for a suitable family of measures.

THEOREM 2. <u>The process</u> $x(t)$ <u>is Markovian iff there is</u> $\beta > 0$ <u>such that</u> $P(v = -\beta) = P(v = \beta) = 1/2$.

In some cases the covariance can be marked out explicitely. In particular, if U is the uniform distribution on $[0,1]$ then an elementary calculation gives the formula

$$E\, x(t)\, x(s) = \frac{3}{4} \min(t,s) + \frac{1}{12} \frac{[\min(t,s)]^2}{\max(t,s)}$$

for $(t,s) \neq (0,0)$. In general the process $x(t)$ has the following structure

$$\sqrt{E|v|}\, W(t) = x(t) + \sqrt{1/2}\, Z(t) \quad ,$$

where $W(t)$ is a standard Brownian motion and $Z(t)$ is a Gaussian process independent of $W(t)$ and given by the following formula:

$$Z(t,\omega) = \int Y(t|v|,\omega)\, P(dv) \ .$$

It is natural to ask whether the trajectories of the limit process $x(t)$ (of Th. 1) are differentiable. It turns out that the answer is negative. With the help of the theorem of Kawada and Kono [7] we can show even a little more. Namely we have.

THEOREM 3. Let $x(t)$ be as above. Then

$$P\left[\varlimsup_{h \searrow 0} \frac{x(t+h) - x(t)}{h} = +\infty \text{ , for each } 0 < t < 1\right] = 1 \ .$$

Let us consider now the model of Harris and Spitzer (see [6],[9], [10]) which may be described as follows:

(S1) $x_0 = 0$;

(S2) $\xi_k = x_k - x_{k-1}$ are exponential random variables with mean one for $k = 0, \pm 1, \pm 2, \ldots$;

(S3) v_k are identically distributed with $E[v_k] = 0$, and $E|v_k| = 1$;

(S4) $\xi_0, v_0, \xi_1, v_1, \xi_{-1}, v_{-1}, \xi_2, \ldots$ are independent random variables.

Here, $\{x_k\}$ is a system of particles in the macroscopic equilibrium (the origin $x_0 = 0$ is included in the system) and what is more, Spitzer proved that if we defined $w_k(t) = (y_k(t) - y(t),\, v_k(t))$ then $\{w_k(t)\}$ would be a random Poisson measure in a phase space for each

$t > 0$. Also for this model Spitzer proved that $Y_A(t) \Longrightarrow W(t)$ as $A \to \infty$, where $Y_A(t)$ is defined as in Theorem 1 and $W(t)$ is a standard Brownian motion.

Now let us consider the model (D) in which $x_k = k$ and $P\{v_k = +1\} = P\{v_k = -1\} = 1/2$. Then, according to Theorem 1, $Y_A(t) \Longrightarrow \sqrt{1/2}\, W(t)$ as $A \to \infty$. On the other hand $y(At)$ may be treated as a symmetric continuous random walk (i.e. $2y(k/2)$ is an ordinary simple random walk), where changes occur at the moments $\pm 1/2, \pm 1, \pm 3/2, \ldots$. The weak convergence is now a consequence of Donsker's theorem on weak convergence of random walks to the Wiener process. Therefore, our model may be viewed as a generalization of Donsker's theorem and this is why we call it "model (D)" . Now, if we specialize in the model (D) the distribution U assuming that it is absolutely continuous, then the assumptions of Dobrushin's theorem are satisfied (see Stone [10], Th.5). Therefore it is plausible that via the macroscopic equilibrium, suggested by Dobrushin's theorem, the corresponding process of the null particle, in this special case, should converge to the Wiener process. This is apparently why Spitzer [9] writes "Now there should be no great difference between a particle system initially on the integers, and one which is initially distributed as a Poisson system. The intuitive idea is in fact supported by the theorems of Dobrushin and Stone". Our theorems on model (D) show that at this point the intuition fails. It is essential that in Dobrushin's theorem the convergence to a Poisson distribution holds only for conditionally compact sets but, if A tends to infinity, the length of the relevant intervals tends to infinity as well. This is the reason why the passage to the limit with $Y_A(t)$, $A \to \infty$, can not be done in two steps: at first passing to the Poisson system and then passing to the Brownian motion.

Now, let us compare the results of Spitzer (model (S)) with the theorems on model (D). Assuming $E|v| = 1$ we get:

(1) $v = \pm 1$,
(S) $Y_A(t) \Longrightarrow W(t)$,
(D) $Y_A(t) \Longrightarrow \sqrt{1/2}\, W(t)$
(2) the distribution of U is symmetric $\neq$ " ± 1 " ,
(S) $Y_A(t) \Longrightarrow W(t)$,
(D) $Y_A(t) \Longrightarrow x(t)$, where $x(t)$ is a <u>non-Markovian</u> Gaussian process.

In both models the assumption $E|v| < \infty$ is essential. In fact, if $E|v| = \infty$, then it is easy to see, that each trajectory $x_k(t)$ collides infinitely many times in each finite time interval.

Proofs.

Proof of the Theorem 1. a) Convergence of finite dimensional distributions. Following Spitzer [8] it suffices to show that:

$$R\left\{A^{-1/2}\left[\sum_{k=1}^{\infty} \chi(x_k+v_kAT < A^{1/2}\ \underline{\alpha}) + \sum_{k=-\infty}^{-1} (\chi[x_k+v_kAT < A^{1/2}\ \underline{\alpha}] - 1)\right] - \underline{\alpha}\right\}$$

$$\to N(0,\ \sigma_{ij})$$

as $A \to \infty$, where $\underline{\alpha} = (\alpha_1,\dots,\alpha_m)$, $T = (t_1,\dots,t_m)$ and $\sigma_{ij} = \min(t_i,t_j)\ E|v| - E\min(t_i|u|,\ t_j|v|)/2$. Because for every $k = 0,\pm1,\pm2,\dots$,

$$A^{-1/2}\ \chi[x_k + v_k\ At_j < A^{1/2}\ \alpha_j] \leqslant A^{-1/2},$$

hence by a classical formulation of the multidimensional central limit theorem it remains to prove that:

$$\text{(i)}\quad A^{-1/2}\ E\left[\sum_{k=1}^{\infty} \chi[x_k+v_kAT < A^{-1/2}\ \underline{\alpha}] + \sum_{k=-\infty}^{-1} (\chi[x_k+v_kAT < A^{1/2}\ \underline{\alpha}]-1)\right] \to \underline{\alpha}$$

as $A \to \infty$, and that for arbitrary $0 \leqslant t_1,\ t_2 \leqslant 1$ and $\alpha_1,\ \alpha_2$:

$$\text{(ii)}\quad A^{-1}\left\{\sum_{k=1}^{\infty}\left[E(\chi[x_k+v_kAt_1 < A^{1/2}\ \alpha_1] - E\ \chi[x_k+v_kAt_1 < A^{1/2}\ \alpha_1])\right.\right.$$

$$\left.\times\ E(\chi\ Ex_k+v_kAt_2 < A^{1/2}\ \alpha_2] - E(\chi[x_k+v_kAt_2 < A^{1/2}\ \alpha_2])\right]$$

$$+ \sum_{k=-1}^{\infty}\left[E[(\chi[x_k+v_kAt_1 < A^{1/2}\alpha_1]-1) - E(\chi\ E[x_k+v_kAt_1 < A^{1/2}\alpha_1]-1)]\right.$$

$$\left.\left.\times\ E[(\chi[x_k+v_kAt_2 < A^{1/2}\alpha_2]-1) - E(\chi[x_k+v_kAt_2 < A^{1/2}\alpha_2]-1)]\right]\right\} \to \sigma_{12}$$

We shall prove (i). We shall use the notation " $g_A \Longleftrightarrow f_A$ " iff $\lim A^{-1/2}\ g_A = \lim A^{-1/2}\ f_A$, $A \to \infty$.

Now, for a uniformly distributed on $[k-1, k)$ random variable s_k ,

$$\sum_{k=1}^{\infty} P(x_k+v_kAt_j < A^{1/2}\alpha_j) - \sum_{k=1}^{\infty} P(x_k+v_kAt_j < -A^{1/2}\alpha_j)$$

$$\Longleftrightarrow \sum_{k=1}^{\infty} P(s_k+v_kAt_j < A^{1/2}\alpha_j) - \sum_{k=1}^{\infty} P(s_k+v_kAt_j < -A^{1/2}\alpha_j)$$

$$\Longleftrightarrow \frac{1}{2}\int_{-\infty}^{\infty}\int_{-\infty}^{\infty} \chi(-A^{1/2}\alpha_j-s,\ A^{1/2}\alpha_j-s)\,ds\,P(dv) = A^{1/2}\alpha_j \quad ,$$

so that the proof of (i) is finished. The proof of (ii) is only a little more complicated and it will be omitted.

b) Weak convergence. Following Billingsley [1], and using the symmetry of our model, we shall show (b) whenever we shall show that for each $\delta > 0$:

$$\lim_{n\to\infty} \limsup_{A\to\infty} \sum_{k=1}^{\infty} P\left[\sup_{t\in[t_k,t_{k+1}]} y(t) - y(t_k) > \sqrt{A}\,\delta/4\right] = 0 \ ,$$

where $t_k = A(k-1)/n$, $k = 1,2,\dots,n$.

It will be clear from the proof that we may set $\delta = 1$ without loss of generality. However,

$$\limsup_{A\to\infty} \sum_{k=1}^{n} P\left[\sup_{t\in[t_k,t_{k+1}]} y(t) - y(t_k) > \sqrt{A}/4\right]$$

$$\leq \limsup_{A\to\infty} \sum_{k=1}^{n} P\left[\sup_{t\in[t_k,t_{k+1}]} y(t) - y(t_k) > \sqrt{A}/4 \wedge y(t_k) < (n^{1/4}-1)\sqrt{A}\right]$$

$$+ \lim_{A\to\infty} \sum_{k=1}^{n} P[y(t_k) \geq (n^{1/4}-1)\sqrt{A}\ ]$$

According to (a) , the second term of the right side tends exponentially to zero while n tends to infinity, We shall show that the first term of the right side in the last expression is $o(n^{-1/4})$ for

A approaching infinity.

Let us consider a motion of our process $Y_A(t)$ on the plane - positions x and times t.

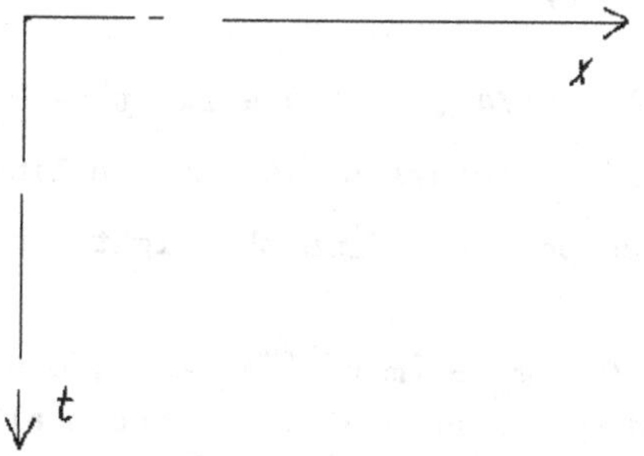

Construct the intervals $I_l(A)$.

$$I_{2l-1}(A) = \left(\frac{1}{8} (l-1) \sqrt{A} \, , \, \frac{1}{8} l \sqrt{A} \right) ,$$

$$I_{2l}(A) = \left(- \frac{1}{8} l \sqrt{A} \, , \, \frac{1}{8} (-l+1) \sqrt{A} \right) ,$$

$l = 1,\dots,L$, where L is the smallest positive integer such that $I_{L+1} \cap (-n^{1/4}A \, , \, n^{1/4}A) = \emptyset$. Then $L \leq 20n^{1/4}$ for large enaugh n. Let $T_{l,k}(A)$ be an orthogonal projection of $I_l(A)$ onto the line $t = t_k$. It is easy to see that for every $k = 1,\dots,n$, $l = 1,2,\dots,L$

$$\lim_{A\to\infty} P \left[K_{l,k}(A) < 10^{-1} \sqrt{A} \right] = 0 ,$$

where $K_{l,k}(A)$ is the number of particles in the interval $I_{l,k}(A)$. Therefore it remains to show that

$$(1) \qquad \limsup_{\to \infty} \sum_{k=1} P \left\{ \sup_{t \in [t_k, t_{k+1}]} y(t) - y(t_k) > \sqrt{A} / 4 \right.$$

$$\left. \wedge \; y(t_k) < (n^{1/4} - 1)\sqrt{A} \wedge K_{y(t_k), y(t_k) + \sqrt{A}/4} > \sqrt{A}/10 \right\} = o(n^{-1/4}) .$$

Following Spitzer [9], for the event in the braces in (1), for each k, we get that

$$\{\dots\} \subset \left\{ \sup_{t\in[t_k,t_{k+1}]} Q_{B_k}(t) > \sqrt{A}/10 \wedge y(t_k) < (n^{1/4}-1)\sqrt{A} \right\},$$

where $B_k(\omega) = y(t_k,\omega) + \sqrt{A}/4$, $Q_{B_k}(t) = L_{B_k}(t) - R_{B_k}(t)$ and $L_{B_k}(t)$ = the number of trajectories hitting the line $x = B_k$ from the left ($R_{B_k}(t)$ - respectively from the right) in the time interval $[t_k,t]$.

Let us put $g_0 = 0$, $g_m = [m\, n^{-1/4}A] + 1$, $m = 1,\dots,M$, $g_{-m} = -g_m$ where $[\cdot]$ is the integer part and M is the smallest positive integer such that $g_{M+1} \notin [0, n^{1/4}A]$. $M < 3\sqrt{n}$ for large enough A. Now, for every $m = 0,\pm1,\dots,\pm M$, $k = 1,\dots,n$ we get easily that

$$\lim_{A\to\infty} P\,[W_{m,k} > \sqrt{A}\,/\,40] = 0\,.$$

($W_{m,k}$ is the number of particles in $[g_{m,k}, g_{m+1},_k]$, and $g_{m,k}$ is the orthogonal projection of g_m onto the line $t = t_k$). Therefore we may assume that for every $m = 0,\pm1,\dots,\pm M$, $k = 1,\dots,n$; $W_{m,k} < \sqrt{A}/40$. Then

$$L_{B_k(\omega)}(t) \leq L_{\overline{m}(\omega),k}(t) + W_{\overline{m}(\omega),k} < L_{\overline{m}(\omega),k}(t) + \sqrt{A}/40$$

$$R_{B_k(\omega)}(t) \geq R_{\overline{m}(\omega),k}(t) - W_{\overline{m}(\omega),k} > R_{\overline{m}(\omega),k}(t) - \sqrt{A}/40\,,$$

where $L_{\overline{m},k}(t)$ ($R_{\overline{m},k}(t)$) is the number of trajectories hitting the line $x = g_m$ in the time interval $[t_k,t]$ from left (right), and $\overline{m}(\omega) = m_i$ iff $m_i \leq B_k(\omega) < m_{i+1}$. To show (1) it is sufficient to check that

$$(2)\quad \limsup_{A\to\infty} \sum_{k=1}^{n} \sum_{m=-M}^{M} P\left[\sup_{t\in[t_k,t_{k+1}]} L_{m,k}(t) - R_{m,k}(t) > \sqrt{A}/20\right]$$

$$< \frac{\text{const}}{\sqrt{n}}\,.$$

But it is easy to check that there exists $c > 0$, such that for every $k = 1,\dots,n$

$$\lim_{A\to\infty} P\,(V(t_{k+1}) > c\cdot \tfrac{A}{n}) = 0$$

where $V(t_{k+1}) = L_{0,k}(t_{k+1}) - R_{0,k}(t_{k+1})$.

Then it remains to prove

$$(3) \qquad \lim_{A\to\infty} \sup 3\sqrt{n} \sum_{k=1} P\Big[\sup_{t\in[t_k,t_{k+1}]} L_{0,k}(t) - R_{0,k}(t) > \sqrt{A}/30$$

$$\wedge \; V(t_{k+1}) \leq c\cdot \tfrac{A}{n}\Big] \leq \frac{\text{const}}{\sqrt{n}} .$$

Hence the following inequality

$$(4) \qquad P\Big[\sup_{t\in[t_k,t_{k+1}]} Q_{0,k}(t) > \sqrt{A}/30 \wedge V(t_{k+1}) \leq c\,\tfrac{A}{n}\Big] \leq \frac{\text{const}}{n^2} ,$$

with universal constant depending only on C , is sufficient to complete the proof.

We shall show that for every $A > 0$, $k = 1,2,\dots,n$; and $N \leq AC/n$

$$P\Big[\sup_{t\in[t_k,t_{k+1}]} Q_{0,k}(t) > \sqrt{A}/30 \wedge V(t_{k+1}) = N|\tau_1,\dots,\tau_N\Big] \leq \frac{\text{const}}{n^2} ,$$

where $(\tau_1,\dots,\tau_N)(\omega)$ are the moments of "collisions" lines $x_k+v_k\cdot t$ with the line $x_k = 0$ in the time interval $[t_k,t_{k+1}]$.

Let us construct random vectors

$$(\tau_1^\nu,\dots,\tau_N^\nu\,(\omega) = (r_1^\nu,\dots,r_N^\nu)$$

on the set

$$\Big\{\omega : (\tau_1,\dots,\tau_N)(\omega) \in [r_1^\nu, r_1^\nu + \tfrac{A}{n}\,2^{-\nu}),\dots,[r_N^\nu, r_N^\nu + \tfrac{A}{n}\,2^{-\nu})\Big\}$$

where $\nu = 1,2,\ldots,r^{\nu}$ are points from the ν-th dyadic partition of the interval $[t_k, t_{k+1}]$.

Now, according to the theorem about convergence of conditional distributions it remains to show that for every positive integer ν :

$$(5) \qquad P\left[{}_{t\in[t_k,t_{k+1}]} Q_{0,k}(t) > \sqrt{A}/30 \wedge V(t_{k+1})\right.$$

$$\left. = N \mid \tau_1^{\nu} = r_1^{\nu},\ldots,\tau_N^{\nu} = r_N^{\nu}\right] \leqslant \frac{\text{const}}{n^2} \quad .$$

But (5) may be deduced without difficulties from the theorem about maxima of partial sums for dependent random variables (cf. Billingsley [1] ,p.75). Thus the proof of Theorem 1 is complete.

Proof of the Theorem 2. Since for $(t,s) \neq (0,0)$ the positive function

$$\text{cov}\,(x(t),x(s)) = \min\,(t,s)\; E|v| - \frac{1}{2}\, E[\,\min\,(t|u|,s|v|)]$$

is symmetric in t and s, it follows that $x(t)$ is Markovian iff

$$\text{cov}\,(x(t),x(s)) = g\,(\min\,(t,s))\cdot h(\max\,(t,s))$$

(cf. Timoszyk [11]). Then for $0 < t \leqslant s \leqslant 1$ we have that $g(t)/g(s) = g(t/s)\cdot C$, for some constant C. Thus $g(t) = t^{-\lambda}\cdot C$, and

$$t\, E|v| - \frac{1}{2}\, E \min\,[t|u|,\, s|v|] = t^{-\lambda}\cdot \text{const} \ .$$

Therefore $-\lambda = 1$, and we have finally that

$$E \min\,[t|u|,\, s|v|] = t\; E\,[\min[\,|u|,\, |v|\,]]$$

for every $0 \leqslant t \leqslant s \leqslant 1$. But this equation may be fulfilled iff $v = \pm\beta$ with probability $1/2$.

Proof of the Theorem 3. According to [7] it is enough to verify that:

$$\text{(6)} \qquad \lim_{h\searrow 0} \sup_{|t-s|\geq h} |E\,(\Delta_h\, x(t)\cdot \Delta_h(x(s))| < \frac{1}{6}$$

where

$$\Delta_h\, x(t) = \frac{x(t+h) - x(t)}{\left\{E[x(t+h) - x(t)]^2\right\}^{1/2}}$$

Let $A_k = [T_0+k\beta\ ,\ T_0+(k+1)\beta]$ $(0,3/2)$, where $k = 0,1,2,\ldots$, $0 < T_0 < 1$, and β, $0 < \beta = \beta(T_0)$, is chosen in such a way that

$$\int_{\substack{T_0/T_0+\beta<|u|/}}\int_{\substack{|v|<(T_0+2\beta)/T_0 \\ |u|\neq|v|}} |v|P(dv)P(du) \leq E|v|/24 .$$

Then (6) is satisfied for $t,s \in A_k$ and for that reason the Theorem 3 holds true for $t,s \in (0,1)$.

References

[1] P.Billingsley, Convergence of probability measures, J.Wiley, New York 1968.

[2] P.Billingsley, Maxima of partial sums, Lecture Notes in Math. vol. 89, Springer-Verlag 1969, 64-76.

[3] Z.Ciesielski, Stochastic systems of particles, this volume, 7-35.

[4] R.Dobrushin, On Poisson laws for distributions of particles in space (in Russian), Ukrain. Mat. Z. 8(1956), 127-134.

[5] R.Gisselquist, A continuum of collision parocess limit theorems, Ann. Prob. 1(1973), 231-239.

[6] T.Harris, Diffusion with collision between particles, J. Appl. Prob. 2(1965), 323-338.

[7] T.Kawada and N.Kono, A remark on nowhere differentiability of sample functions of Gaussian processes, Proc. Jap. Acad. 47(1971), 932-934.

[8] F.Spitzer, Uniform motion with elastic collision of an infinite particle system, J.Math. Mech. 18(1969), 973-990.

[9] F.Spitzer, Random processes defined through the interaction of an infinite particle system, Lecture Notes in Math. vol. 89, Springer-Verlag 1969, 201-223.

[10] C.Stone, On a theorem by Dobrushin, Ann. Math. Stat. 39 1968), 1391-1401.

[11] W.Tımoszyk, A characterization of Gaussian processes that are Markovian, Coll. Math. 3(1974), 157-167,

WINTER SCHOOL
IN PROBABILITY
Karpacz 1975
Springer's LNM

472

EXTREME POINT METHOD IN PROBABILITY THEORY (*)

By K. Urbanik
Wrocław University
and
the Institute of Mathematics, Polish Academy of Sciences

Introduction.

The famous Krein-Milman Theorem ([15], [21]), generalized by Choquet ([3]), asserts that every point of a compact convex set K in a locally convex Hausdorff space is the barycenter of a probability measure concentrated on the closure $\overline{e(K)}$ of the set $e(K)$ of extreme points of K. More precisely, for every point $x \in K$ there exists a representing probability measure m supported by $\overline{e(K)}$ such that

$$f(x) = \int_{\overline{e(K)}} f(y)\, m(dy) \tag{1}$$

for every continuous linear functional f. Further, for metrizable sets K it was proved by Choquet ([3]) that there exists a representing measure m in (1) concentrated on the set $e(K)$ of extreme points of K. This Theorem has frequently been employed to classical characterization theorems for measures and functions (see, for instance,

*) A part of this paper was presented on the Summer School on Probability and Mathematical Statistics organized by the Institute of Mathematics and Mechanics, Bulgarian Academy of Sciences, Varna, May 10-24, 1974.

[21], Section 2). In 1963 Kendall ([10]) obtained an integral representation for the Laplace-Stieltjes transform of infinitely divisible probability measures on the positive half-line. The method of his proof consisted in finding the extreme points of a certain compact convex set K formed by the logarithm l of the Laplace-Stieltjes transform of infinitely divisible distributions satisfying the condition $l(1) = -1$. The set K is a base of the cone of the logarithm of the Laplace-Stieltjes transform of arbitrary infinitely divisible probability distribution on the positive half-line. The Krein-Milman-Choquet Theorem applied to the set K followed by exponentianting leads to the formula

$$\int_0^\infty e^{-ux} P(dx) = \exp\left(-\int_0^\infty \frac{1-e^{-ux}}{1-e^{-x}} \lambda(dx)\right)$$

where λ is a finite Borel measure on the positive half-line. This interesting result stimulated further applications of the extreme point method in probability theory. In 1966 Johansen used the same idea and obtained a version of the well-known Lévy-Khinchine representation of the characteristic function of infinitely divisible probability distributions on the real line ([9]). It is not my intention to write a complete systematic and book-like account or to present a key tour through the highlights of extreme point method in probability theory. What I intent is an extended hike through a series of topics especially connected with limit problems in probability theory.

Section 1 contains a rather systematic presentation of results in the one-dimensional case. In other Sections I would like to present general ideas and only to sketch what the full results are. In what follows I shall consider as an underlying linear locally convex space the space of Borel signed countably additive measures over a compact metric space equipped with the weak topology. In this case the set of all probability measures is compact and as a set K I shall always choose an appropriate closed and convex its subset. The problem will consist in finding the extreme points of the set K in question.

1. Lévy's probability measures on the real line.

It is well-known that infinitely divisible probability measures are limiting distributions of sums

$$\sum_{k=1}^{k_n} X_{nk} - a_n$$

where in each row $X_{n1}, X_{n2}, \ldots, X_{nk_n}$ the random variables are independent and uniformly asymptotically negligible, i.e.

$$\lim_{n\to\infty} \max_{1\leq k\leq k_n} P(|X_{nk}| \geq c) = 0$$

for every positive number c and, moreover, a_n are suitably chosen constants. Moreover, the characteristic function of infinitely divisible probability measures is given by the Lévy-Khinchine formula

$$(2) \qquad \varphi(t) = \exp\left\{iat + \int_{-\infty}^{\infty} \left(e^{itx} - 1 - \frac{itx}{1+x^2}\right) \frac{1+x^2}{x^2}\, \mu(dx)\right\} ,$$

where a is a real constant and the spectral measure μ is a Borel measure on the real line.

Let $X_1, X_2, \ldots$ be a sequence of independent not necessarily identically distributed random variables. Consider normed sums

$$(3) \qquad a_n \sum_{k=1}^{n} X_k - b_n$$

where a_n are positive numbers, b_n are real numbers and the triangular array $X_{nk} = a_n X_k$ $(k = 1,2,\ldots,n)$ consists of uniformly asymptotically negligible random variables. It is clear that the limit distribution of (3) is infinitely divisible provided it exists. The limit measures are called Lévy's measures. P. Lévy established a necessary and sufficient condition for μ to be a spectral measure of a Lévy's probability measure. The Lévy's Theorem can be formulated as follows: a measure μ is the spectral measure of a Lévy's measure if and only if the function $f(x) = \mu(-\infty,x)$ has left and right derivatives denoted by F' invariably and

$$\frac{1+x^2}{x} F'(x)$$

do not increase on the half-lines $(-\infty,0)$ and $(0,\infty)$ respectively

([17]). In 1968 I proved, using the extreme point method, another representation formula for Lévy's probability measures which can be regarded as an analoque of the Lévy-Khinchine formula ([27]).

We start from the well-known properties of Lévy's measures. We say that a probability measure with the characteristic function φ is self-decomposable whenever for every number c from the interval $(0,1)$ there exists the characteristic function φ_c of a random variable such that

$$\varphi(t) = \varphi(ct)\,\varphi_c(t) \qquad (-\infty < t < \infty) .$$

One can prove that φ_c is the characteristic function of an infinitely divisible law. Furthermore, a probability measure is a Lévy's measure if and only if it is self-decomposable (see [18], p. 323). Let μ_c denote the spectral measure corresponding to the probability distribution with the characteristic function φ_c. By a simple calculation we get the formula

$$\mu_c(E) = \mu(E) - c^2 \int_{c^{-1}E} \frac{1 + x^2}{1 + c^2x^2}\, \mu(dx) , \tag{4}$$

where E is an arbitrary Borel subset of the real line and $c^{-1}E = \left\{c^{-1}x : x \in E\right\}$. Hence we get the following criterion.

PROPOSITION 1. μ __is the spectral measure of a Lévy's measure if, and only if for every__ $c \in (0,1)$ __the measure defined by the right -hand side of__ (4) __is non-negative__.

Let M be the set of all finite Borel measures μ defined on the compactified real line $[-\infty,\infty]$ for which the corresponding measures μ_c $(c \in (0,1))$ are non-negative. Of course, by Proposition 1, the subset of M consisting of measures concentrated on the real line $(-\infty,\infty)$ coincides with the set of all spectral measures corresponding to Lévy's probability measures. From proposition 1 it follows that the set M is convex and closed under the weak convergence. Let K be the subset of M consisting of probability measures. It is evident tha the set K is a compact convex base of the cone M. Our aim is to find all extreme points of K.

A subset F of the compactified line said to be invariant if for every number c from $(0,1)$ the inclusion $cF \subset F$ holds. Given an invariant set F and a measure μ, we infer that for any $c \in (0,1)$ the measure defined by (4) fulfils the condition

$$(\mu|F)_c = \mu_c|F \quad ,$$

where $\mu|F$ and $\mu_c|F$ denote the restriction of μ and μ_c to F respectively. Consequently, for any invariant set F we have $\mu|F \in M$ provided $\mu \in M$. Hence we get the following

PROPOSITION 2. <u>The extreme points of</u> K <u>are concentrated on one of the folowing sets</u>: $\{-\infty\}$, $\{0\}$, $\{\infty\}$, $(-\infty,0)$, $(0,\infty)$.

It is obvious that the extreme points of K supported by one--point sets $\{-\infty\}$, $\{0\}$, $\{\infty\}$ are degenerate probability measures $\delta_{-\infty}$, δ_0 and δ_∞ respectively. Moreover, the point mapping $x \to -x$ induces a one-to-one correspondence between extreme points of K supported by the half-lines $(-\infty,0)$ and $(0,\infty)$ respectively. Consequently, it suffices to find extreme points which are probability measures concentrated on the half-line $(-\infty,0)$.

We proceed now to the investigation of these extreme points. For any probability measure μ supported by $(-\infty,0)$ we put

$$J_\mu(u) = \int_{-\infty}^{-e^{-u}} \frac{1+v^2}{v^2}\,\mu(dv) \qquad (-\infty < u < \infty) . \tag{5}$$

Given $a < b$ and $t > 0$, we have, by a simple calculation,

$$J_\mu(b) - J_\mu(a) - J_\mu(b-t) + J_\mu(a-t) = \int_{[-e^{-a},-e^{-b}]} \frac{1+v^2}{v^2}\,\mu_c(dv) ,$$

where $c = e^{-t}$. Hence, by Proposition 1, we conclude that the probability measure μ supported by $(-\infty,0)$ belongs to K if and only if

$$J_\mu(b) - J_\mu(a) - J_\mu(b-t) + J_\mu(a-t) \geq 0 \tag{6}$$

for all $t > 0$ and $a < b$. Substituting $b = a + t$ into (6) we get the inequality

$$J_\mu(a) \leq \frac{1}{2}\,(J_\mu(a-t) + J_\mu(a+t))$$

which shows that the function J_μ is convex. Since it is also monotone non-decreasing with $J_\mu(-\infty) = 0$, we have the representation

$$J_\mu(t) = \int_{-\infty}^{t} q_\mu(u)\,du \quad ,$$

where q_μ is non-negative and monotone non-decreasing. Of course, without loss of generality we may assume that q_μ is continuous from the left. In this case the function q_μ is uniquely determined by μ . Conversely, μ is also uniquely determined by q_μ . In fact, by (5), for any Borel subset E of $(-\infty,0)$ we have the formula

$$\mu(E) = \int_E q_\mu(-\log|x|)\,\frac{|x|}{1+x^2}\,dx \quad .$$

In particular,

$$\int_{-\infty}^{0} q_\mu(-\log|x|)\,\frac{|x|}{1+x^2}\,dx = 1 \quad .$$

Suppose we have a non-negative, continuous from the left monotone non -decreasing function q normalized by the condition

$$\int_{-\infty}^{0} q\,(-\log|x|)\,\frac{|x|}{1+x^2}\,dx = 1 \tag{7}$$

Setting for $E \subset (-\infty,0)$

$$\mu(E) = \int_E q\,(-\log|x|)\,\frac{|x|}{1+x^2}\,dx \quad , \tag{8}$$

we get a probability measure concentrated on the half-line $(-\infty,0)$. By (5) $J_\mu(u) = \int_{-\infty}^{u} q(v)\,dv$ and, consequently, for $t > 0$ and $a < b$

$$J_\mu(b) - J_\mu(a) - J_\mu(b-t) + J_\mu(a-t) = \int_a^b (q(v) - q(v-t))\,dv \geq 0$$

which proves inequality (6). Thus μ belongs to K. In other words the formula (8) establishes a one-to-one correspondence between all measures μ from K concentrared on the half-line $(-\infty,0)$ and all non-negative continuous from the left monotone non-decreasing functions q normalized by the condition (7). It is obvious that the correspondence in question preserves convex combinations of elements. Consequently, extreme points of K are transformed into extreme points.

In order to determine all functions q being extreme points we note that functions which for some s are not constant on both half-lines $(-\infty,s]$ and (s,∞) are not extreme points because they admit a decomposition $q = cq_1 + (1-c)\, q_2$, where

$$0 < c = \int_{-\infty}^{-e^{-s}} q(-\log|x|)\, \frac{|x|}{1+x^2}\, dx + q(s) \int_{-e^{-s}}^{0} \frac{|x|}{1+x^2}\, dx < 1$$

and

$$q_1(x) = c^{-1} \min\, (q(x)\, ,\, q(s))$$

$$q_2(x) = (1-c)^{-1} \max\, (q(x) - q(s)\, ,\, 0)\, .$$

Thus the extreme points are functions which for some s are constant on both half-lines $(-\infty,s]$ and (s,∞). Hence and from (7) and (8), by a simple calculation, we obtain all extreme points of K concentrated on $(-\infty,0)$. They depend upon a parameter $z \in (-\infty,0)$ and are given by the formula

$$(9) \qquad \lambda_z(E) = \frac{2}{\log\,(1+z^2)} \int_E \chi_z(u)\, \frac{|u|}{1+u^2}\, du\, ,$$

where $E \subset (-\infty,0)$ and χ_z is the indicator of the interval $(z,0)$. Applying the mapping $x \to -x$ we get remaining extreme points concentrated on the half-line $(0,\infty)$. They are given for $E \subset (0,\infty)$ by the same formula (9), where $z \in (0,\infty)$ and χ_z is the indicator of the interval $(0,z)$. Thus denoting by $\lambda_{-\infty}$, λ_0, λ_∞ the measures $\delta_{-\infty}$, δ_0, δ_∞ respectively we get the following

PROPOSITION 3. $e(K) = \left\{\lambda_z : z \in [-\infty,\infty]\right\}$ <u>and the mapping</u>

$\lambda_z \to z$ *is a homeomorphism between* $e(K)$ *and* $[-\infty,\infty]$.

Now we can apply Krein-Milman-Choquet Theorem. We then get that for every measure $\mu \in K$ there exists a probability measure λ on $[-\infty,\infty]$ such that for all continuous functions f on $[-\infty,\infty]$ we have

$$\int_{[-\infty,\infty]} f(y)\,\mu(dy) = \int_{[-\infty,\infty]} \Big(\int_{[-\infty,\infty]} f(y)\,\lambda_z(dy)\Big)\,\lambda(dz)\,.$$

Moreover, the measure λ will assign zero mass to the set $\{-\infty\}$ $\{\infty\}$ if, and only if μ does so. Hence we get the following statement: μ is the spectral measure corresponding to a Lévy's probability measure if and only if there exists a finite Borel measure ν on the real line such that

$$(10) \qquad \int_{\infty}^{\infty} f(y)\,\mu(dy) \quad \frac{1}{2}\int_{-\infty}^{\infty} \Big(\int_{-\infty}^{\infty} f(y)\,\lambda_z(dy)\Big)\,\nu(dz)$$

for all continuous bounded functions f on $(-\infty,\infty)$. The coefficient $\frac{1}{2}$ is taken for simplicity of further calculations. Namely, taking in (10) as a function f the Lévy-Khinchine kernel from (2) we get the following characterization theorem.

THEOREM 1. *A function* φ *is the characteristic function of a Lévy's probability measure on the real line if and only if*

$$\varphi(t) = \exp\left\{ i\gamma t + \int_{-\infty}^{\infty} \Big(\int_0^{tu} \frac{e^{iv}-1}{v}\,dv - it \arctan u\Big) \frac{\nu(du)}{\log(1+u^2)} \right\},$$

where γ *is a real constant,* ν *is a finite Borel measure on* $(-\infty,\infty)$ *and the integrand is defined as its limiting value* $-\frac{1}{4}t^2$ *when* $u=0$.

One can prove that the function φ determines γ and ν uniquely (see [27]).

2. *Lévy's probability measures on Euclidean spaces.*

This section is concerned with a description of limit laws arising, roughly speaking, from affine modification of the partial sums of a sequence of independent R^N - valued random variables. The main results were published in [28]. The limit laws in the case of a sequence of independent and identically distributed random variables, i.e. the operator-stable probability measures on R^N were considered by Sharpe in [24].

In the sequel by $\lambda * \mu$ we shall denote the convolution of two probability measures λ and μ. We call a probability measure on R^N full if its support is not contained in any (N-1) - dimensional hyperplane of R^N. Let End R^N denote the semigroup of all linear operators in R^N with the composition as a semigroup operation. Further, let Aut R^N denote the group of all non-singular linear operators in R^N. For any $A \in$ End R^N and any probability measure λ on R^N let $A\lambda$ denote the measure defined by the formula $A\lambda(E) = \lambda(A^{-1}(E))$ for all Borel subsets E of R^N. It is easy to check that if φ is the characteristic function of λ, then $\varphi(A^*t)$ $(t \in R^N)$, where A denotes the adjoint operator, is the characteristic function of $A\lambda$.

Now we shall introduce the concept of operator-decomposability of measures which will play the same role in the multivariate case as the concept of self-decomposability in the one-dimensional case. We denote by $E(\lambda)$ the set of all operators $A \in$ End R^N for which the equation $\lambda = A\lambda * \lambda_A$ holds for a certain probability measure λ_A. Further, by $A(\lambda)$ we denote the subset of $E(\lambda)$ consisting of those operators A for which we may take $\lambda_A = \delta_a$ for some element $a \in R^N$. Here δ_a denotes the probability measure concentrated at the point a. For full measures λ one can prove that $E(\lambda)$ is a compact subsemigroup of End R^N and $A(\lambda)$ is a compact subgroup of Aut R^N (see [28], p. 121).

I believe that the knowledge of the algebraic and topological structure of $E(\lambda)$ and $A(\lambda)$ may enrich the theory of multivariate probability distributions and be of some mathematical interest. We quote only the following statement: Let D be a symmetric positive operator on R^N. Suppose that $E(\lambda)$ consists of all operators A for which the operator $D - ADA^*$ is non-negative. Then λ is a Gaussian probability measure.

A detailed investigation of $E(\lambda)$ and $A(\lambda)$ for full probability measures λ is given in [28], Chapter 1. As an example we quote the following result. In what follows for any operator $A \in$ End R^N det A will denote the determinant of the matrix representation of A with respect to an orthonormal basis in R^N.

PROPOSITION 4. <u>Given a full probability measure</u> λ <u>on</u> R^N, <u>if</u> $A \in E(\lambda)$ <u>and</u> $|\det A| = 1$, <u>then</u> $A \in A(\lambda)$.

Proof. Consider the monothetic compact subsemigroup S of $E(\lambda)$ generated by the operator A. By a Theorem of Numakura (see [19], [20], p. 109) the limit points of the sequence $\{A^n\}$ form a group G which is the minimal ideal of S and S contains exactly one idempotent, namely the unit J_0 of G. Of course, $J_0 \in E(\lambda)$ and $\det J_0 = 1$. Further. from the equation $\lambda = J_0\lambda * \lambda_{J_0}$ we get the following one $J_0\lambda = J_0\lambda * J_0\lambda_{J_0}$. Hence it follows that the characteristic function of the measure $J_0 \lambda_{J_0}$ is equal to 1 in a neighbourhood of the origin. Since J_0 is non-singular, we infer that $\lambda_{J_0} = \delta_0$. Consequently,

$$\lambda = J_0\lambda \tag{11}$$

Further, there exists $B \in G$ such that $BA = AB = J_0$. Of course, $|\det B| = 1$ and $B \in E(\lambda)$. From the equations

$$\lambda = A\lambda * \lambda_A \quad , \qquad\qquad \lambda = B\lambda * \lambda_B$$

we get the equation

$$\lambda = BA\,\lambda * \lambda_B * B\,\lambda_A \quad ,$$

which, by (11), implies

$$\lambda = \lambda * \lambda_B * B\,\lambda_A \quad .$$

Hence it follows that the absolute value of the characteristic function of the probability measure $B\lambda_A$ is equal to 1 in a neighbourhood of the origin. Thus $B\lambda_A$ and, consequently, λ_A is concentrated at a single point which completes the proof.

Now we proceed to the main problem we study. Suppose that $\{X_n\}$ is a sequence of independent R^N - valued random variables and assume that $\{A_n\}$ and $\{a_n\}$ are sequences from Aut R^N and R^N respectively such that the triangular array A_nX_k $(k = 1,2,\ldots,n;\ n = 1,2,\ldots)$ consists of uniformly asymptotically negligible random variables and the distribution of

$$A_n \sum_{k=1}^{n} X_k + a_n$$

converges to a measure μ ; what can be said about the limit measure μ ? Converting this to a problem involving only measures we ask which measures μ can arise as limits of sequences $A_n(\mu_1 * \mu_2 * \dots * \mu_n) * \delta_{a_n}$ where $\{\mu_n\}$ is an arbitrary sequence of probability measures in R^N, such that $A_n \mu_k$ $(k = 1,2,\dots,n;\ n = 1,2,\dots)$ form a uniformly infinitesimal triangular array. The limit measures μ will be called Lévy's measures. The sequence $\{A_n\}$ of operators will be called a norming sequence. It is easy to prove the following statements:

PROPOSITION 5. For every norming sequence $\{A_n\}$ corresponding to a full measure the relation $\lim_{n\to\infty} A_n = 0$ holds. Moreover, one can always choose a norming sequence $\{A_n\}$ with the property

$$\lim_{n\to\infty} A_{n+1} A_n^{-1} = 1$$

(see [28], p. 126-127).

We say that a probability measure λ is Q - decomposable, if $E(\lambda)$ contains a one-parameter semigroup $\exp tQ$ $(t \geq 0)$. The following statement is an analoque of the self-decomposability condition in the one-dimensional case.

PROPOSITION 6. A full probability measure on R^N is a Lévy's measure if, and only if it is Q - decomposable for an operator Q with eigenvalues having a negative real part.

Our next aim is to give a representation of the characteristic function of Q - decomposable probability measures. First we introduce some auxiliary spaces.

Let S^m be the m -dimensional unit sphere and $[-\infty,\infty]$ the compactified real line. Put $H^N = S^{N-1} \times [-\infty,\infty]$. Obviously, the space H^N is compact. We define a congruence relation in H^N as follows: $\langle x,t\rangle \sim \langle y,u\rangle$ where $x,y \in S^{N-1}$ and $t,u \in [-\infty,\infty]$ if and only if there exists a real number s such that $\exp sQx = y$ and $u = t+s$. Suppose that $\langle x_n,t_n\rangle \sim \langle y_n,u_n\rangle$ $(n = 1,2,\dots)$ and the sequences $\{\langle x_n,t_n\rangle\}$ and $\{\langle y_n,u_n\rangle\}$ converge to $\langle x,t\rangle$ and $\langle y,u\rangle$ respectively. Then for some real numbers s_n $\exp s_n Q x_n = y_n$ $(n = 1,2,\dots)$. Since all eigenvalues of Q have negative real part, the last equations and the compactness of S^{N-1} imply that the sequence $\{s_n\}$ is bounded.

If s is its limit point, then $\exp sQx = y$ and $u = t+s$. Thus $\langle x,t\rangle \sim \langle y,u\rangle$ and, consequently, the quotient space $H^N/_\sim$ is compact (see [2], p. 97). The element of $H^N/_\sim$, i.e. the equivalence class containing $\langle x,t\rangle$ from H^N will be denoted by $[x,t]$. We define a one-parameter group T_s $(-\infty < s < \infty)$ of transformations of $H^N/_\sim$ by assuming

$$T_s\,[x,t] = [x, s+t].$$

Since $\lim_{t\to\infty} \exp tQ = 0$ and for every $z \in R^N \setminus \{0\}$ $\lim_{t\to-\infty} \|\exp tQz\| = \infty$, each element $z \in R^N \setminus \{0\}$ can be represented in the form $z = \exp tQx$ where $x \in S^{N-1}$ and $-\infty < t < \infty$. In general this representation is not unique. But $z = \exp uQy$, where $y \in S^{N-1}$ and $-\infty < u < \infty$ if and only if $\langle x,t\rangle \sim \langle y,u\rangle$. Thus the mapping $\exp tQx \to [x,t]$ is an embedding of $R^N \setminus \{0\}$ into $H^N/_\sim$.

Suppose we have a Q - decomposable probability measure on R^N. Obviously, it is infinitely divisible and, consequently, its characteristic function φ can be written in the Lévy-Khinchine form

$$\varphi(y) = \exp\left\{ i(a,y) - \frac{1}{2}(Dy,y) + \int_{R^N\setminus\{0\}} \left(e^{i(y,u)} - 1 - \frac{i(y,u)}{1+\|u\|^2}\right) \frac{1+\|u\|^2}{\|u\|^2}\, \mu(du) \right\},$$

where a is a vector from R^N, D is a symmetric non-negative operator in R^N and the spectral measure μ is a finite Borel measure on $R^N \setminus \{0\}$. Since $R^N \setminus \{0\}$ can be embedded into $H^N/_\sim$, the spectral measures μ can be identify with finite Borel measure on $H^N/_\sim$ concentrated on $R^N \setminus \{0\}$. Futher, introducing the notation $|[x,t]| = \|\exp tQx\|$ if $-\infty < t < \infty$, $|[x,\infty]| = 0$ and $|[x,-\infty]| = \infty$, we put for every finite measure μ on $H^N/_\sim$ and every real number t

$$\mu_t(E) = \mu(E) - \int_{T_{-t}E} \frac{|T_t v|^2\,(1+|v|^2)}{(1+|T_t v|^2)\,|v|^2}\,\mu(dv) .$$

Let M_N be the set of all finite measures μ on $H^N/_\sim$ for which μ_t is non-negative for all t. It is clear that M_N is convex and closed. Further, we note that for Q - decomposable measures λ with the spectral measure μ, μ_t is the spectral measure of the probability distribution λ_t defined by the equation $\lambda = \exp tQ\lambda * \lambda_t$. Moreover, we have the following

PROPOSITION 7. <u>Spectral measures of Q - decomposable probability distributions coincide with measures from M_N concentrated on $R^N \setminus \{0\}$.</u>

Let K_N be the subset of M_N consisting of probability measures. Of course, K_N is convex and compact. Our aim is to find its extreme points. In the same way as in the one-dimensional case we prove that the restriction of any measure from M_N to a T_t - invariant $(-\infty < t < \infty)$ subset of $H^N/_\sim$ belongs to M_N too. Hence it follows that extreme points of K_N are concentrated on orbits of elements of $H^N/_\sim$. More precisely, we have an analogue of Proposition 2.

PROPOSITION 8. <u>The extreme points of K_N are concentrated on one of the following sets: $\{[x,-\infty]\}$, $\{[x,\infty]\}$ $\{[x,t] : -\infty < t < \infty\}$ where $x \in S^{N-1}$.</u>

The extreme points supported by one-point sets $\{[x,-\infty]\}$, $\{[x,\infty]\}$ $(x \in S^{N-1})$ are degenerate probability distributions. It remains to describe extreme points concentrated on each orbit $F_x = \{[x,t] : -\infty < t < \infty\}$ $(x \in S^{N-1})$. One can easily prove that the homeomorphism $[x,t] \to e^t$ from F_x onto the half-line $(0,\infty)$ establishes a one-to-one correspondence between measures from K_N concentrated on F_x and measures from K concentrated on $(0,\infty)$. Thus Proposition 3 yields

PROPOSITION 9. <u>The set of extreme points of K_N concentrated on F_x consists of all probability measures $\lambda_{(x,a)}$ $(-\infty < a < \infty)$ defined by means of the formula</u>

$$\int_{F_x} f(z)\,\lambda_{(x,a)}(dz) = C_a \int_a^\infty f([x,u])\, \frac{\|[x,u]\|^2}{1+\|[x,u]\|^2}\, du \quad ,$$

<u>where</u>

$$C_a^{-1} = \int_a^\infty \frac{\|[x,u]\|^2}{1+\|[x,u]\|^2}\, du \quad .$$

Now the Krein-Milman-Choquet Theorem yields an integral representation of spectral measures corresponding to Q - decomposable measures. Finally, we get the representation of the characteristic function of Q - decomposable probability measures.

THEOREM 2. _Suppose that all eigenvalues of Q have negative real part. A function φ on R^N is the characteristic function of a Q - decomposable probability measure if and only if_

$$\varphi(y) = \exp\left\{ i(a,y) - \frac{1}{2}(Dy,y) \right.$$

$$\left. + \int_{R^N \setminus \{0\}} \int_0^\infty \left(e^{i(y,\exp tQx)} - 1 - \frac{i(y,\exp tQx)}{1+\|\exp tQx\|^2}\right) dt \, \frac{\nu(dx)}{\log(1+\|x\|^2)} \right\},$$

where a is a vector from R^N, D is a symmetric non-negative operator, $QD + DQ^*$ is non-positive and ν is a finite Borel measure on $R^N \setminus \{0\}$. Moreover, the function φ determines the triplet a,D,ν uniquely.

3. Generalized convolutions.

It has been pointed out by Kingman [14] that we can study random walks with sphrical symmetry wholly in terms of the lengths of the vectors concerned, the use of the ordinary characteristic function being replaced by that of the radial characteristic function with a Bessel function as the kernel. It is natural to try to generalize this procedure. In [25] I introduced the concept of generalized convolutions on probability measures on the positive half-line possesing appropriate analogues of the most important properties of ordinary convolution. The theory of generalized convolutions was developed in [1], [16] and [26].

We summarize briefly those parts of this theory which we shall need later. Let Π be the class of all probability measures on $[0,\infty)$, endowed with the weak convergence. We write δ_x for the unit mass at the point $x \geq 0$, and write τ_x for the map given by

$$(\tau_x P)(E) = P(x^{-1} E)$$

for $P \in \Pi$ and E a Borel subset of $[0,\infty)$. Let $\circ$ be a Π - valued commutative and associative binary operation on Π which is linear on convex combinations, continuous, has δ_0 as unit element, is homogeneous under the maps τ_x $(x \geq 0)$ and satisfies the following axiom, called the law of large numbers for point measures: there exists a sequence of positive numbers c_n such that the sequence of measures $\tau_{c_n} S_1^{\circ n}$ is convergent to some measure other than δ_0. Here $P^{\circ n}$ denotes the n-th power of P under the operation $\circ$. Such a map $\circ$ is called a generalized convolution. $(\Pi,\circ)$ is then a generalized convolution algebra.

Except ordinary convolution the most important non-trivial example of generalized convolutions is provided by Kingman's work [14] on spherically symmetric random walks:

(12) $$(P \circ Q)(E) = \frac{1}{2} \int_0^\infty \int_0^\infty [\chi_E(x\, y) + \chi_E(|x-y|)]\, P(dx)Q(dy) ,$$

and, for $\beta > 1$,

(13) $$(P \circ Q)(E)$$

$$= \frac{\Gamma(\frac{\beta}{2})}{\Gamma(\frac{\beta-1}{2})\sqrt{\pi}} \int_0^\infty \int_0^\infty \int_{-1}^1 \chi_E((x^2+y^2+2xyz)^{\frac{1}{2}})(1-z^2)^{\frac{\beta-3}{2}} dz\, P(dx)Q(dy) .$$

Another example of a generalized convolution is given by the formula

(14) $$(P \circ Q)(E) = \int_0^\infty \int_0^\infty \chi_E\, (\max\,(x,y))\, P(dx)Q(dy) .$$

Recently, a non-standard example of a generalized convolution was given in [16] :

(15) $$(P \circ Q)(E) = \int_0^\infty \int_0^\infty \left(1 - \frac{\min(x,y)}{\max(x,y)}\right)^n \chi_E(\text{amx}\,(x,y))P(dx)Q(dy)$$

$$+ \sum_{k=0}^{n} \binom{n}{k}^2 \int_0^\infty \int_0^\infty x^{n-k}\, y^k \int_{\max(x,y)} \chi_E(z)\left(1 - \frac{x}{z}\right)^k \left(1 - \frac{y}{z}\right)^{n-k}$$

$$\times\ z^{-1-n}\left(\frac{kx}{z-x}+\frac{(n-k)y}{z-y}\right)\,dz\,P(dx)Q(dy)\ .$$

Here χ_E denotes the indicator of the Borel set E and $\min(0,0)/\max(0,0)$ is assumed to be 0 .

A real-valued continuous map h defined on Π which is linear on convex combinations and multiplicative with respect to $\circ$ is called a homomorphism of Π . Each generalized convolution algebra admits two trivial homomorphisms $h \equiv 0$ and $h \equiv 1$. Algebras admitting a non-trivial homomorphism are called regular. It is easy to verify that the algebra with the convolution (14) is not regular and all remaining algebras with the convolution (12), (13) and (15) are regular.

Let Φ_P be a one-to-one map $P \to \Phi_P$ between Π and the class of real-valued functions defined on $[0,\infty)$. If Φ_P is linear on convex combinations, multiplicative under $\circ$ and is such that weak convergence on Π is equivalent to uniform convergence on compacta, Φ_P is said to be a characteristic function on Π . It is possible to prove that many of the familiar properties of the ordinary charcteristic function are shared by the characteristic function of a generalized convolution algebra. Among many results we record the following ones: Φ_P is given by the integral transform

$$\Phi_P(t) = \int_0^\infty \Omega(tx)\,P(dx)\ .$$

In particular, for generalized convolution (12), (13) and (15) we have $\Omega(x) = \cos x$,

$$\Omega(x) = \Gamma\left(\frac{\beta}{2}\right)\left(\frac{2}{x}\right)^{\frac{\beta}{2}-1} J_{\frac{\beta}{2}-1}(x)\quad ,$$

where J_ν is the Bessel function and $\Omega(x) = (1-x)^n$ for $0 \leq x \leq 1$ and $\Omega(x) = 0$ for $x > 1$ respectively. Moreover, one can prove that a generalized convolution algebra admits a characteristic function if and only if it is regular. Further $\Omega(x) < 1$ for all sufficiently small positive x . We define $w(x) = 1 - \Omega(x)$ if $0 \leq x \leq x_0$ and $w(x) = 1 - \Omega(x_0)$ if $x > x_0$, where x_0 is a positive number such that $\Omega(x) < 1$ whenever $0 < x \leq x_0$.

Infinite divisibility may be defined as usual with respect to the semigroup operation $\circ$. We have the characterization: Φ_p is the characteristic function of an infinitely divisible measure in $(\Pi,\circ)$ if and only if

(16) $$\Phi_p(t) = \exp \int_0^\infty \frac{\Omega(tx) - 1}{w(x)}\, m(dx)$$

for some finite Borel measure m on $[0,\infty)$.

One may define a triangular array

$$\left\{ P_{n,k} : 1 \leq k \leq k_n \; ; \; n \geq 1 \right\}$$

of measures in $(\Pi,\circ)$ just as in the clasical case. An array is convergent to P if $P = \lim_{n\to\infty} P_{n1} \circ P_{n2} \circ \ldots \circ P_{nk_n}$. It was proved in [25] that a measure P is infinitely divisible if and only if it is the limit of a triangular array $\{P_{nk}\}$, where the measures P_{nk} are uniformly asymptotically negligible in the usual sense.

The problem we study is enunciated as follows: suppose we have a sequence $P_1, P_2, \ldots$ of probability measures from Π and a sequence $c_1, c_2, \ldots$ of positive numbers such that $\tau_{c_n} P_k$ $(k = 1,2,\ldots,n$; $n = 1,2,\ldots$) are uniformly asymptotically negligible. We ask which measures P can arise as limits of sequence $\tau_{c_n}(P_1 \circ P_2 \circ \ldots \circ P_n)$ $(n = 1,2,\ldots$) ? For ordinary convolution we get Lévy's measures on $[0,\infty)$. In general the class of all possible limit probability measures will be called Lévy's measures in the generalized convolution algebra $(\Pi,\circ)$. We call a measure P self-decomposable in $(\Pi,\circ)$ if for every number c satisfying the condition $0 < c < 1$ there exists a measure $Q_c \in \Pi$ such that $P = \tau_c P \circ Q_c$. One can prove the following

PROPOSITION 10. The class of self-decomposable measures in $(\Pi,\circ)$ coincides with the class of limit distributions of sequences $\tau_{c_n}(P_1 \circ P_2 \circ \ldots \circ P_n)$ $(n = 1,2,\ldots$) where $\tau_{c_n} P_k$ $(k = 1,2,\ldots,n$; $n = 1,2,\ldots$) are uniformly asymptotically negligible.

We proceed now to a representation problem for the characteristic function of self-decomposable measures in a generalized convolution algebra.

A subset E of the compactified half-line $[0,\infty]$ is said to be separated from the origin if its closure is contained in $(0,\infty]$. Let m be a finite Borel measure on $[0,\infty]$. For any Borel subset E of $[0,\infty]$ separated from the origin we put

$$T_m(E) = \int_E \frac{m(dx)}{w(x)} \quad ,$$

where the integrand is assumed to be $(1-\Omega(x_0))^{-1}$ if $x=\infty$. Denote by M the set of all finite Borel measures m on $[0,\infty]$ satisfying for all numbers c $(0 < c < 1)$ and all Borel subsets E separated from the origin the following condition

$$I_m(E) - I_m(c^{-1}E) \geq 0 \; . \tag{17}$$

It is clear that the set M is convex and closed. Moreover, one can prove the subset of M consisting of measures concentrated on $[0,\infty)$ coincides with the set of all spectral measures m in (16) corresponding to self-decomposable measures. Let K be the subset of M consisting of probability measures. Of course, K is a convex compact set. Analogously, as in the case of ordinary convolution algebra, we can prove that extreme points of K are measure concentrated on one of the following sets: $\{0\}$, $\{\infty\}$ and $(0,\infty)$. Our aim is to find all extreme points concentrated on $(0,\infty)$. Put

$$J_m(x) = \int_x^\infty \frac{m(du)}{w(u)} \qquad (x > 0) \; .$$

Obviously, $I_m([a,b)) = J_m(a) - J_m(b)$. It is easy to see that $m \in M$ if and only if (17) holds for all c $(0 < c < 1)$ and all subsets E of the form $[a,b)$. Consequently, $m \in M$ if for every triplet a,b,c satisfying the conditions $0 < c < 1$, $0 < a < b$ the inequality

$$J_m(a) - J_m(b) - J_m(\tfrac{a}{c}) + J_m(\tfrac{b}{c}) \geq 0 \tag{18}$$

is true. Introducing the notation

$$F(x) = J_m(e^x) \qquad (-\infty < x < \infty) \; .$$

and substituting $a = e^{x-h}$, $b = e^x$, $c = e^{-h}$ $(-\infty < x < \infty$, $0 < h < \infty)$ into (18) we get the inequality

$$F(x) \leq \frac{1}{2} (F(x-h) + F(x+h)) \quad .$$

Thus the function F is convex on the real line. Moreover, it is also monotone non-increasing with $F(\infty) = 0$. Consequently, it can be represented in the form

$$F(x) = \int_x^\infty q_m(u)\, du \quad ,$$

where q_m is continuous from the left, monotone non-increasing and non-negative. Further,

$$m(E) = \int_E w(x)\, q_m(\log x) \frac{dx}{x} \quad .$$

The mapping $m \to q_m$ establishes a one-to-one correspondence between the set K and the set of all continuous from the left, monotone non -increasing and non-negative functions q satisfying the condition

$$\int_0^\infty w(x)\, q(\log x) \frac{dx}{x} = 1 \; .$$

It is obvious that the correspondence in question preserves convex combinations of elements. Consequently, extreme points of K concentrated on $(0,\infty)$ are transformed into extreme points and conversely. This enables us to prove that the measures from $e(K)$ concentrated on $(0,\infty)$ coincides with the measures m_x $(0 < x < \infty)$ defined by means of the formula

$$m_x(E) = c(x) \int_\varepsilon \chi_x(u) \frac{du}{u}$$

where χ_E is the indicator of the interval $(0,x)$ and

$$c(x)^{-1} = \int_0^x \frac{w(u)}{u}\, du \; .$$

Hence by virtue of the Krein-Milman-Choquet Theorem we get the representation of self-decomposable measures.

THEOREM 3. The class of characteristic function of self-decomposable measures in $(\Pi, \circ)$ coincides with the class of all functions of the form

$$\Phi_P(t) = \exp \int_0^\infty \int_0^x \frac{\Omega(tu) - 1}{w(u)} du \left(\int_0^x \frac{w(v)}{v} dv \right)^{-1} \nu(dx)$$

where ν is a finite Borel measure on $[0,\infty)$.

We apply this result to the random walk with spherical symmetry in the Euclidean space R^N. Namely, consider a random walk in R^N given by

$$S_n = X_1 + X_2 + \dots + X_n \qquad (n = 1,2,\dots)$$

where $X_1, X_2, \dots$ are independent random vectors in R^N having spherical symmetric distribution. The probability distribution of $\|S_n\|$ is the generalized convolution of distributions of $\|X_1\|$, $\|X_2\|$, ... defined by formula (12) for $N = 1$ and (13) for $N > 1$ with $\beta = N$. The asymptotic behaviour of $\|S_n\|$ $(n = 1,2,\dots)$ can be described in terms of the limit distribution of the sequence $c_n \|S_n\|$ $(n = 1,2,\dots)$ where c_n are suitable chosen positive numbers. It is clear that the class of all possible limit distributions coincides with the class of all self-decomposable probability measures in the generalized convolution algebra in question. Since

$$\int_0^x \frac{w(u)}{u} du \sim \log(1 + x^2)$$

on the whole positive half-line, we get, by virtue of Theorem 3, the following statement: the class of all possible limit distributions of sequences $c_n\|S_n\|$, where $c_n > 0$ and $S_n = X_1 + X_2 + \dots + X_n$ $(n = 1,2,\dots)$, $\{X_k\}$ being independent random vectors from R^N with spherical symmetric distribution coincides with the class of all probability distribution P on $[0,\infty)$ whose integral transformation

$$\Phi_p(t) = \int_0^\infty \Gamma\left(\frac{N}{2}\right)\left(\frac{2}{tx}\right)^{\frac{N}{2}-1} J_{\frac{N}{2}-1}(tx)\, P(dx)$$

is of the form

$$\Phi_p(t) = \exp \int_0^\infty \int_0^x \frac{\Gamma\left(\frac{N}{2}\right)\left(\frac{2}{tu}\right)^{\frac{N}{2}-1} J_{\frac{N}{2}-1}(tu) - 1}{u}\, du\, \frac{\nu(dx)}{\log(1+x^2)} ,$$

where ν is a finite Borel measure on $[0,\infty)$.

4. Exchangeable random variables.

A sequence $X_1, X_2, \ldots$ of random variables is called a sequence of exchangeable random variables whenever for every system of distinct integers $i_1, i_2, \ldots i_n$ the distribution of $X_{i_1}, X_{i_2}, \ldots, X_{i_n}$ depends only on n and does not depend upon a special choice of $i_1, i_2, \ldots, i_n$. Suppose that $X_1, X_2, \ldots$ assume their values from a metric compact space, say A . Then the probability distribution of $X_1, X_2, \ldots$ induces the probability measure μ on the catesian product $B = A \times A \times \ldots$ This measure has the following property: let T be an arbitrary one-to-one mapping of the positive integers onto themselves leaving all but a finite number of integers fixed. Then μ is invariant under T . Obviously, this property characterizes measures μ induced by exchangeable random variables.Jules Haag seems to have been the first autor to discuss sequences of exangeable random variables [7]. This paper deals only with two-valued random variables. It hints at, but does not rigorously state or prove, the representation theorem for this case. Somewhat later, exchangeable two-valued random variables were independently introduced by de Finetti who proved a representation theorem (see [5] and [6]). This case has also been trated by Khinchine ([12], [13]). De Finetti also proved the theorem for the case of real-valued random variables. This case has also been treated by Dynkin [4]. Dynkin's technique also applies to random variables on all spaces that are, in a certain sense separable. Another proof of Dynkin's result was given by Ryll-Nardzewski [23]. We mention also the proof given by Rényi and Révész ([22]) based on the concept of stable sequences of random events and a simple martingale proof due to Kendall ([11]). In

1955 Hewitt and Savage presented a proof based on the Krein-Milman -Choquet Theorem. Now we shall sketch their idea of proof.

It is clear the set K of all probability measures μ induced by an exchangeable sequence of A - valued random variables on the Cartesuan product $B = A \times A \times \ldots$ with compact space A is convex and compact. In order to find all its extreme points we prove some auxiliary propositions.

Given a system $E_1, E_2, \ldots, E_n$ of Borel subsets of A we denote by $C(E_1, E_2, \ldots, E_n)$ the cylinder $E_1 \times E_2 \times \ldots \times E_n \times A \times A \times \ldots$

PROPOSITION 11. _Let_ $\mu \in K$ _and let_ $E_1, E_2, \ldots, E_n$ _be a system of Borel subsets of_ A . _Then_

$$\mu C(E_1, E_2, \ldots, E_n, E_1, E_2, \ldots, E_n) \geq [\mu C(E_1, E_2, \ldots, E_n)]^2 . \tag{19}$$

Proof. Let the cylinders appearing on the left and right sides of (19) be denoted by C_1 and C_2 , respectively, and let χ_r $(r = 1,2,\ldots)$ be the indicator of the cylinder

$$C(\underbrace{A, A, \ldots, A}_{(r-1)n \text{ times}}, E_1, E_2, \ldots, E_n) .$$

Then, for every positive integer k , it is easy to see that

$$\int_B \left(\sum_{r=1}^{k} \chi_r(b) \right) \mu(db) = k \, \mu C_2 . \tag{20}$$

Furthermore, a similar direct calculation shows that

$$\int_B \left(\sum_{r=1}^{k} \chi_r(b) \right)^2 \mu(db) \tag{21}$$

$$= \sum_{r=1}^{k} \sum_{s=1}^{k} \int_B \chi_r(b) \, \chi_s(b) \, \mu(db) = k \, \mu C_2 + k(k-1) \, \mu C_1 .$$

Applying the Cauchy-Schwarz inequality, we have

(22) $$\left(\int_B \left(\sum_{r=1}^{k} X_r(b)\right) \mu(db)\right)^2 \leq \int_B \left(\sum_{r=1}^{k} X_r(b)\right)^2 \mu(db) .$$

Combining (20), (21) and (22), we find

$$\mu C_1 \geq (\mu C_2)^2 - \frac{1}{k}(\mu C_2 - \mu C_1) .$$

Since the last inequality holds for all positive integers k, our proposition is proved.

PROPOSITION 12. Let μ be an element of K such that equality obtains in (19) for all positive integers n and all Borel sets $E_1, E_2, \dots, E_n$. Then μ is an extreme point of K.

Proof. Suppose the contrary, i.e. $\mu = \alpha\mu_1 + (1-\alpha)\mu_2$, where $\mu_1, \mu_2 \in K$, $\mu_1 \neq \mu_2$ and $0 < \alpha < 1$. Since all measures from K are determined by their values on cylinders, there exists a cylinder $C_2 = C(E_1, E_2, \dots, E_n)$ such that $\mu_1 C_2 \neq \mu_2 C_2$. Let

$$C_1 = C(E_1, E_2, \dots, E_n, E_1, E_2, \dots, E_n) .$$

Then, in view of Proposition 11, we have

(23) $$\mu C_1 = \alpha\mu_1 C_1 + (1-\alpha)\mu_2 C_1 \geq \alpha(\mu_1 C_2)^2 + (1-\alpha)(\mu_2 C_2)^2 .$$

Applying the Cauchy-Schwarz inequality again, we have

(24) $$(\alpha\mu_1 C_2 + (1-\alpha)\mu_2 C_2)^2 < \alpha(\mu_1 C_2)^2 + (1-\alpha)(\mu_2 C_2)^2 ,$$

since the conditions for equality obviously fail here. Combining (23) and (24), we obtain

$$\mu C_1 > (\alpha\mu_1 C_2 + (1-\alpha)\mu_2 C_2)^2 = (\mu C_2)^2 ,$$

so that strict inequality holds in (19). This proves our Proposition.

Let Π_A be the set of all Borel probability measures on A. It is obvious, by Proposition 12, that for every $p \in \Pi_A$ the product measure $p \times p \times \dots$ is an extreme point of K . The converse implication is also true. Namely, we have

PROPOSITION 13. $e(K) = \{p \times p \times \dots : p \in \Pi_A\}$.

Proof. To prove our Proposition it suffices to show that each extreme point of K is a product measure. Suppose the contrary. Then must exist Borel subsets E , $F_1, F_2, \dots, F_n$ of A such that

$$\mu C(E \, , F_1, F_2, \dots, F_n) \neq \mu C(E)\, \mu C(F_1, F_2, \dots, F_n) \; . \tag{25}$$

Obviously, $0 < \mu C(E) < 1$. We define two measures from K by means of the formulae

$$\mu_1 \, C(E_1, E_2, \dots, E_n) = \frac{\mu C(E, E_1, E_2, \dots, E_n)}{\mu C(E)}$$

$$\mu_2 \, C(E_1, E_2, \dots, E_n) = \frac{\mu C(A \quad E, \, E_1, E_2, \dots, E_n)}{\mu C(A \quad E)} \; .$$

Since

$$\mu C(E_1, E_2, \dots, E_n) = \mu C(A \, , E_1, E_2, \dots, E_n)$$

$$= \mu C(E \, , E_1, E_2, \dots, E_n) + \mu C(A \quad E \, , E_1, E_2, \dots, E_n)$$

$$= \alpha \mu_1 C(E_1, E_2, \dots, E_n) + (1-\alpha)\, \mu_2 C(E_1, E_2, \dots, E_n)$$

with $\alpha = \mu C(E)$, we infer, by (25), that $\mu_1 \neq \mu_2$ which contradicts the assumption that μ is an extreme point of K . The Proposition is thus proved.

Applying the Krein-Milman-Choquet Theorem we get de Finetti's result.

THEOREM 4. <u>Each exchangeable probability measure</u> μ <u>on the Cartesian product</u> $A \times A \times \dots$ <u>with a compact</u> A <u>is of the form</u>

$$\mu = \int_{\Pi_A} (p \times p \times \ldots)\, \lambda(dp),$$

where λ is a probability measure on Π_A .

References

[1] N.H. Bingham, Factorization theory and domains of attraction for generalized convolution algebras, Proceedinds of the London Math. Soc. 23 (1971), 16-30 .

[2] N. Bourbaki, Éléments de mathématique, I, Les structures fondamentales de l'analyse, Livre III, Topologie générale, Paris 1951 .

[3] G. Choquet, Le théorème do représentation intégrale dans les ensembles convexes compacts, ANN. Inst. Fourier, Grenoble 10 (1960), 333-344.

[4] E.B. Dynkin, Klassy ekvivalentnych slučainych veličin, Uspehi Mat. Nauk. 8, vol 54 (1953), 125-134 .

[5] B. de Finetti, Funzione caratteristica di un fenomeno aleatorio, Atti della R. Accademia Nazionale dei Lincei, Ser. 6, Memorie, Classe di Scienze Fisiche, Matematiche e Naturali, 4 (1931), 251-299 .

[6] B. de Finetti, La prévision: ses lois logiques, ses sources subjectives, Ann. Inst. Henri Poincaré 7(1937), 1-68 .

[7] J.Haag, Sur un problème général do probabilités et ses diverses applications, Proceedings of the International Congress of Mathematicians, Toronto, 1924, 659-674.

[8] E.Hewitt and L.J.Savage, Symmetric measures on cartesian products, Trans. Amer. Math. Soc. 80 (1955), 470-501 .

[9] S.Johansen, An application of extreme point methods to the representation of infinitely divisible distributions, Zeitschrift für Wahrscheinlichkeitstheorie und verw. Gebiete 5 (1966), 304-316 .

[10] D.G.Kendall, Extreme - point methods in stochastic analysis, Zeitschrift für Warscheinlichkeitstheorie und verw. Gebiete, 1 (1963), 295-300 .

[11] D.G.Kendall, On finite and infinite sequences of exchangeable events, Studia Scientiarum Math. Hung. 2 (1967), 319-327 .

[12] A.Ya.Khinchine, O klassah ekvivalentnyh sobytii, Doklady Akad. Nauk SSSR, 85 (1952), 713-714 .

[13] A.Ya.Khinchine, Sur les classes d'événements équivalents, Mat. Sbornik 39 (1932), 40-43 .

[14] J.F.Kingman, Random walks with spherical symmetry, Colloquium on Combinatorical Methods in Probability Theory, Aarhus (1962), 40-46 .

[15] M.G.Krein and D.P.Milman, On extreme points of regularly convex sets, Studia Math. 9 (1940), 133-138 .

[16] J.Kucharczak and K.Urbanik, Quasi-stable functions, Bull. Acad. Pol. Sci., Série des sci. math., astronom. et phys. 22 (1974), 263-268 .

[17] P.Lévy, Théorie de l'addition des variables aléatoires, Paris 1954 .

[18] M.Loève, Probability theory, New York, 1950 .

[19] K.Numakura, On bicompact semigroups, Math. J. Okayama Univ. 1 (1952), 99-108 .

[20] A.B.Paalman-de Miranda, Topological semigroups, Amsterdam 1964 .

[21] R.P.Phelps, Lectures on Choquet's theorem, Princeton 1966 .

[22] A.Rényi and P.Révész, A study of sequences of equivalent events as special stable sequences, Publ. Math. Debrecen, 10 (1963), 319-325 .

[23] Cz.Ryll-Nardzewski, On stationary sequences of random variables and the de Finetti's equivalence, Coll. Math. 4 (1955), 149 -156 .

[24] M.Sharpe, Operator - stable probability distributions on vector groups, Trans. Amer. Math. Soc. 136 (1969), 51-65 .

[25] K.Urbanik, Generalized convolutions, Studia Math. 23 (1964) 217-245 .

[26] K.Urbanik, Generalized convolutions II, Studia Math. 45 (1973), 57-70 .

[27] K.Urbanik, A representation of self-decomposable distributions, Bull. Acad. Pol. Sci., Série des sci. math.,astronom. et phys. 16 (1968), 196-204 .

[28] K.Urbanik, Lévy's probability measures on Euclidean spaces, Studia Math. 44 (1972), 119-148 .

WINTER SCHOOL
ON PROBABILITY
Karpacz 1975
Springer's LNM

472

STABLE SYMMETRIC PROBABILITY LAWS IN QUANTUM MECHANICS

By K.Urbanik
Wrocław University
and
the Institute of Mathematics, Polish Academy of Sciences

The aim of this note is to find all possible symmetric limit probability distribution operators for arbitrarily normed sequences of sums of independent pairs of canonical observables. The main, rather unexpected, result is that the ground states are the only such limit operators.

The paper has its origin in the quantum-mechanical central limit theorem proved by C.D.Cushen and R.L.Hudson in [3] for sequences of pairs of canonical observables. Let p,q be self-adjoint not necessarily bouded operators acting in a separable Hilbert space H. They constitute a pair of canonical observables if there exists a dense linear manifold D in H contained in the domains of definition of p,q and invariant under p,q such that p,q restricted to D satisfy the Heisenberg commutation relation $pq - qp = -i\hbar 1$, where 1 is the identity operator and, moreover, the operator $p^2 + q^2$ on D is essentially self-adjoint. J.Dixmier proved in [4] that p,q constitute a pair of cannonical observables if and only if there exists a Hilbert space H' and a unitary mapping

$$T_{p,q} : H \to L^2(R) \otimes H' \tag{1}$$

form H onto the Hilbert space tensor product with H' of the Hilbert space $L^2(R)$ of Lebesque square-integrable functions on the real line R, such that for $x,y \in R$

$$\exp ixp = T^{-1}_{p,q} (\exp ixp_0 \otimes 1_{H'}) T_{p,q}$$

$$\exp iyq = T^{-1}_{p,q} (\exp iyq_0 \otimes 1_{H'}) T_{p,q}$$

where p_0, q_0 are the Schrödinger operators in $L^2(R)$, i.e. the unique self-adjoint extensions of the operators defined initially on the manifold of infinitely differentiable rapidly decreasing functions by

$$p_0 \, \psi(t) = -i \hbar \frac{d\psi(t)}{dt} ,$$

$$q_0 \, \psi(t) = t \, \psi(t) .$$

Now, let ρ be a density operator in H. C.D.Cushen and R.L.Hudson introduced in [3] the concept of the distribution operator $\rho_{p,q}$ for the canonical observables p,q in the state ρ as follows. Corresponding to the decomposition (1) of H there exists a reduced density operator for ρ in $L^2(R)$, which may be defined as the unique density operator $\rho_{p,q}$ in $L^2(R)$ such that for an arbitrary bounded operator A in $L^2(R)$ we have the equality

$$\mathrm{tr} A \rho_{p,q} = \mathrm{tr}\, T^{-1}_{p,q} (A \otimes 1_{H'}) T_{p,q} \rho$$

Let J be the unitary operator in $L^2(R)$ defined by the formula

$$(J\psi)(t) = \psi(-t) .$$

We say that the observables p,q are symmetrically distributed at the state ρ if the density operator $\rho_{p,q}$ in $L^2(R)$ satisfies the conditions

$$J \rho_{p,q} = \rho_{p,q} J = \rho_{p,q} .$$

It is evident that in this case the eigen-functions of $\rho_{p,q}$ corresponding to a positive eigen-value are even functions. Thus the density operator of a symmetric probability distribution can be written in the form

(2)
$$\rho_{p,q} = \sum_{k=1}^{\infty} \lambda_k \Pi_{\psi_k} ,$$

where Π_{ψ_k} denotes the projector onto the one-dimensional subspace spanned by ψ_k and the functions $\psi_1, \psi_2, \ldots$ form an orthonormal system of even functions. Obviously, $\lambda_k = 0$ and $\sum_{k=1}^{\infty} \lambda_k = 1$. Moreover, it is easy to verify that p,q and $-p,-q$ are identically distributed whenever p,q are symmetrically distributed.

Now let the Hilbert space H carry a sequence $p_1,q_1,p_2,q_2,\ldots$ of pairs of canonical observables. We asume that the pairs p_j,q_j $(j = 1,2,\ldots)$ are physically independent in the sense that the corresponding Weyl operators $\exp i(x\, p_j + y\, q_j)$ $(j = 1,2,\ldots\, ;\ x,y \in R)$ commute with one another. Then for any integer n there exists a Hilbert space H_n' and a unitary mapping

$$T_{p_1,q_1,\ldots,p_n,q_n} : H \to L^2(R)^{\otimes n} \otimes H_n' ,$$

where the power is taken in the sense of the tensor product, such that for $x,y \in R$ and $j = 1,2,\ldots,n$

$$\exp ixp_j = T^{-1}_{p_1,q_1,\ldots,p_nq_n} (1_{L^2(R)^{\otimes(j-1)}} \otimes \exp ixp_0$$
$$\otimes 1_{L^2(R)^{\otimes(n-j)}} \otimes 1_{H_n'})\, T_{p_1,q_1,\ldots,p_n,q_n} ,$$

and

$$\exp iyq_j = T^{-1}_{p_1,q_1,\ldots,p_n,q_n} (1_{L^2(R)^{\otimes(j-1)}} \otimes \exp iyq_0$$
$$\otimes 1_{L^2(R)^{\otimes(n-j)}} \otimes 1_{H_n'})\, T_{p_1,q_1,\ldots,p_n,q_n}$$

where p_0, q_0 are the Schrödinger operators in $L^2(R)$.

The joint distribution operator $\rho_{p_1,q_1,\ldots,p_n,q_n}$ in the state ρ is the reduced density operator for ρ in $L^2(R)^{\otimes n}$ defined by the formula

$$\operatorname{tr} B\, \rho_{p_1,q_1,\ldots,p_n,q_n}$$

$$= \operatorname{tr} T^{-1}_{p_1,q_1,\ldots,p_n,q_n} (B \otimes 1_{H_n'})\, T_{p_1,q_1,\ldots,p_n,q_n}$$

for any bounded operator B in $L^2(R)^{\otimes n}$.

The sequence of pairs of canonical observables $p_1,q_1,p_2,q_2,\ldots$ is said to be stochastically independent in the state ρ if for any integer n

$$\rho_{p_1,q_1,\ldots,p_n q_n} = \rho_{p_1,q_1} \otimes \rho_{p_2,q_2} \otimes \cdots \otimes \rho_{p_n,q_n}$$

Further, the pairs $p_1,q_1,p_2,q_2,\ldots$ are said to be identically distributed in the state ρ if the corresponding distribution operators ρ_{p_n,q_n} are identical.

Suppose we have a sequence $p_1,q_1,p_2,q_2,\ldots$ of physically and stochastically independent and symmetrically distributed canonical observables. Given a norming sequence $a_1,a_2,\ldots$ of positive numbers we construct a new sequence $\overline{p}_1,\overline{q}_1,\overline{p}_2,\overline{q}_2,\ldots$ of observables by assuming

$$\overline{p}_n = a_n(p_1 + p_2 + \cdots + p_n) ,$$

$$\overline{q}_n = a_n(q_1 + q_2 + \cdots + q_n)$$

or more precisely through Stone's Theorem as the infinitesimal generators of continuous one-parameter groups:

$$\exp ix\overline{p}_n = \prod_{j=1}^{n} \exp ixa_n p_j ,$$

$$\exp iy\bar{q}_j = \prod_{j=1}^{n} \exp iya_n q_j .$$

Let $\rho_{\bar{p}_n,\bar{q}_n}$ be the distribution operator of the pair $\bar{p}_n,\bar{q}_n$ at the state ρ. We say that the operator ρ_0 defined in $L^2(R)$ is the limit distribution operator of the sequence $\bar{p}_1,\bar{q}_1,\bar{p}_2,\bar{q}_2,\ldots$ if for every bounded operator A in $L^2(R)$ the relation

$$\lim_{n\to\infty} \operatorname{tr} A\, \rho_{\bar{p}_n,\bar{q}_n} = \operatorname{tr} A\, \rho_0$$

holds. It is clear that the limit distribution operator is uniquely determined. Accordingly the classical probability theory the limit distribution operator ρ_0 will be called stable and symmetric. The aim of this paper is to find all stable symmetric probability distribution operators.

A probability distribution operator is said to be a ground state if and only if it is the projector onto the span of a pure state for which in the Heisenberg inequality the equality holds. It was proved in [6] that a symmetric pure state ψ is a ground state if and only if

$$\psi(t) = b \sqrt[4]{\frac{2a}{\pi}}\; e^{-at^2}$$

where $|b| = 1$, and $a > 0$, and the equality holds almost everywhere in the sense of the Lebesque measure in R.

THEOREM. __A symmetric probability distribution operator is stable if and only if it is a ground state.__

__Proof__. The proof of the Theorem is based on the concept of quasi-characteristic function which is a specialisation to the case of a single degree of freedom of the vacuum expectation functional for a representation of the canonical commutation relations for infinitely many degrees of freedom, arising in quantum field theory (see [1]). According to [3] the function $\operatorname{tr} \exp i(xp + yq)\rho$ $(x,y \in R)$ is the quasi-characteristic function for the pair of canonical observables p,q in the state ρ. There is a simple relationship between the quasi-characteristic function for p,q and the linear combinations $xp + yq$

$(x,y \in R)$. Namely

$$\operatorname{tr} \exp i(xp + yq)\rho = \int_{-\infty}^{\infty} e^{iu} \operatorname{tr} E_{xp+yq}(du)\rho ,$$

where E_{xp+yq} is the spectral measure associated with the operator $xp+yq$. Let $f_n(x,y)$ be the quasi-characteristic function of the pair $\bar{p}_n, \bar{q}_n$ at the state ρ , i.e.

$$f_n(x,y) = \operatorname{tr} \exp i(x\bar{p}_n + y\bar{q}_n)\rho .$$

Of course, without loss of generality we may assume that $a_1 = 1$. Then, by the stochastic independence of pairs p_j, q_j $(j = 1,2,\ldots)$

$$f_n(x,y) = f_1(a_n x , a_n y)^n \qquad (n = 1,2,\ldots) .$$

Let f be quasi-characteristic function of the limit distribution operator. Then

$$(3) \qquad f(x,y) = \lim_{n\to\infty} f_n(a_n x , a_n y)^n \qquad (x,y \in R)$$

By the Proposition 5 in [3] the functions f^2 and f_1^2 are the characteristic functions of some classical probability distributions on R^2 . By virtue of (3) we infer that f^2 is the characteristic function of a classical stable and symmetric probability distribution on R^2 . Thus, by the Lévy's Theorem ([5], p. 221-224).

$$f^2(x,y) = \exp\left\{ - \int_{u^2+v^2=1} |xu + yv|^p \, m_0(du,dv)\right\} ,$$

where $0 < p \leq 2$ and m_0 is a finite Borel measure on the unit circle Consequently, by the continuity of f and the initial condition $f(0,0) = 1$ we get the formula

$$(4) \qquad f(x,y) = \exp\left\{ - \int_{u^2+v^2=1} |xu + yv|^p \, m(du,dv)\right\} ,$$

where $0 < p \leq 2$ and m is a finite Borel measure in the unit circle. On the other hand taking into account Szilard-Wigner expression for the joint density function of p_o, q_o ([6], [7]) at a pure state we get the formula

$$(5) \qquad f(x,y) = \operatorname{tr} \exp i(xp_o + yq_o)\rho_o$$

$$= \sum_{k=1}^{\infty} \lambda_k \int_{-\infty}^{\infty} e^{ixz} \psi_k(z + \frac{yh}{2}) \psi_k^*(z - \frac{yh}{2}) dz \quad ,$$

where $\rho_o = \sum_{k=1}^{\infty} \lambda_k \Pi_{\psi_k}$, ψ^* is the complex conjugate of ψ and the eigenfunctions ψ_k are even . Moreover, the function f is Lebesgue square-integrable over R^2 (see [3], Proposition 6). Hence it follows that

$$(6) \qquad \int_{u^2+v^2=1} |xu + yv|^p \, m(du,dv) > 0$$

whenever $x^2 + y^2 > 0$.
In fact, the equality

$$\int_{u^2+v^2=1} |x_o u + y_o v|^p \, m(du,dv) = 0$$

for $x_o^2 + y_o^2 > 0$ would imply that the measure m is concentrated on a two-point set $\{\langle u_o, v_o\rangle , \langle -u_o, -v_o\rangle\}$. Then, by [4],

$$f(x,y) = \exp \left\{ -c \, |xu_o + yv_o|^p \right\} ,$$

where $c > 0$. But this function is not square-integrable over R^2 , which shows that (6) is always true.

Let us introduce the notation

$$(7) \qquad h(x) = \int_{u^2+v^2=1} |u + x^{-1} v|^p \, m(du,dv) \qquad \text{for } x \neq 0$$

By (6) we have the inequalities

(8) $$a = \inf_{x \neq 0} h(x) > 0$$

and

(9) $$b = \inf_{x \neq 0} |x|^p h(x) > 0$$

Consequenthy, by (4),

$$f(x,y) \leq \exp\left\{- |x|^p h\left(\frac{x}{y}\right)\right\} \leq \exp\left\{- a |x|^p\right\} ,$$

which shows that for every y the function $f(x,y)$ is Lebesgue integrable over R . Thus the function

$$F(z,y) = \frac{1}{2\pi} \int_{-\infty}^{\infty} e^{-ixz} f(x,y)\, dx$$

is continuous in z and y .
By (5)

$$F(z,y) = \sum_{k=1}^{\infty} \lambda_k \phi_k \left(z + \frac{yh}{2}\right) \phi_k^* \left(z - \frac{yh}{2}\right)$$

for every y and almost every z . Since the eigen-functions ϕ_k are even, we have also the equality

$$F^* \left(\frac{yh}{2} , \frac{2z}{h}\right) = \sum_{k=1}^{\infty} \lambda_k \phi_k \left(z + \frac{yh}{2}\right) \phi_k^* \left(z - \frac{yh}{2}\right)$$

for every z and almost every y . Hence it follows that

(10) $$F(z,y) = F^* \left(\frac{yh}{2} , \frac{2z}{h}\right)$$

for almost every pair y,z . By virtue of the continuity of F consequently, (10) holds for all y and z and,

$$F(z,0) = F^{*}\left(0 , \frac{2z}{h}\right) \tag{11}$$

for $z \in R$. By (4)

$$F(z,0) = \frac{1}{2\pi} \int_{-\infty}^{\infty} \exp\left(- ixz - c |x|^{p}\right) dx ,$$

where, according to (6), $c = \int\limits_{u^2+v^2=1} |u|^p \, m(du,dv) > 0$. In other words, the function $F(z,0)$ is the density function of the symmetric stable law with the exponent p on the real line. Taking into account Bergström's results on the asymptotic behaviour of this density function we have the relation

$$\lim_{z\to\infty} z^{1+p} F(z,0) = \frac{c}{\pi} \Gamma(p + 1) \sin \frac{\pi p}{2} . \tag{12}$$

Now consider the function $F(0,z)$. From (4) and (7) it follows that

$$F(0,z) = \frac{z}{2\pi} \int_{-\infty}^{\infty} \exp\left(- |z|^{p} |x|^{p} h(x)\right) dx . \tag{13}$$

Consequently, by (8) and (9),

$$\int_{|x|<1} \exp\left(- |z|^{p} |x|^{p} h(x)\right) dx \leq 2 \exp\left(- b |z|^{p}\right)$$

and, for $a|z|^{p} > p^{-1}$

$$\int_{|x|\geq 1} \exp\left(- |z|^{p} |x|^{p} h(x)\right) dx \leq 2 |z| \exp\left(- a |z|^{p}\right) .$$

Hence and from (11) we get the relation

$$\lim_{z\to\infty} z^{1+p} F(0,z) = 0 .$$

Comparying this with (11) and (12) we infer that $p = 2$.

Consequently, by (4) and (6),

$$f(x,y) = \exp\,(-\,c_1x^2 - c_2y^2 - c_3xy)$$

where $c_1 > 0$ and $c_2 > 0$. Consequently,

$$F(z,y) = \frac{1}{\sqrt{\pi\, c_1}} \exp\left\{ -\frac{z^2}{c_1} - i\,\frac{c_3}{2c_1}\, yz - \left(c_2 - \frac{c_3^2}{4c_1}\right) y^2 \right\}$$

which, by (10), implies $c_3 = 0$ and $c_1c_2 = \frac{\hbar^2}{4}$ the latter being essentially the Heisenberg uncertainty principle. Thus

$$f(x,y) = \operatorname{tr} \exp i(xp_0 + yq_0)\, \Pi_\varphi$$

where φ is the ground state defined by the formula

$$\varphi(t) = (2\pi c_1)^{-\frac{1}{4}} \exp\left(-\frac{t^2}{4c_1}\right).$$

Consequently, by (5) and the one-to-one correspondence between the quasi-characteristic functions and probability distribution operators we get the equality $\rho_0 = \Pi_\varphi$ which completes the proof that each stable and symmetric probability distribution operator is a ground state.

Now we shall prove that each symmetric ground state is stable. Given a ground state

$$\text{(12)} \qquad \varphi(t) = \sqrt[4]{\frac{2a}{\pi}}\; e^{-at^2},$$

where $a > 0$, we consider the Cartesian product $R \times R \times \ldots$ of countably many real lines with the product measure $\mu = \nu \times \nu \times \ldots$ where ν is the Gaussian probability measure on R with mean 0 and the variance $(4a)^{-1}$. Let H be the Hilbert space of μ square-integrable functions on $R \times R \times \ldots$. We define a sequence of physically independent pairs of canonical observables $p_1, q_1, p_2, q_2, \ldots$ by setting

$$p_j\, \psi(t_1,t_2,\ldots) = -\,i\hbar \frac{\partial}{\partial t_j}\, \psi(t_1,t_2,\ldots) - 4\, ai\hbar t_j\, \psi(t_1,t_2,\ldots),$$

$$q_j\, \psi(t_1,t_2,\dots) = t_j\, \psi(t_1,t_2,\dots)$$

for $j = 1,2,\dots$. Further, as the state ρ we take the projector onto the span of the function identically equal to 1 . It is clear that the correspondence between functions of n variables

$$\psi(t_1,t_2,\dots,t_n) \to \psi(t_1,t_2,\dots,t_n)\, \left(\frac{2a}{\pi}\right)^{\frac{n}{4}} \exp\left(- a \sum_{j=1}^{n} t_j^2\right)$$

can be extended to a unitary mapping

$$T_{p_1,q_1,\dots,p_n,q_n} : H \to L^2(R)^{\otimes n} \otimes H_n' \; .$$

Hence, by a simple calculation, we get the formula

$$\rho_{p_1,q_1,\dots,p_n,q_n} = \rho_{p_1,q_1} \otimes \rho_{p_2,q_2} \otimes \dots \otimes \rho_{p_n,q_n}$$

where, for every $j = 1,2,\dots,n$, ρ_{p_j,q_j} is the projector onto the span of the function (12). Thus the pairs of observables $p_1,q_1,p_2,q_2,\dots$ are stochastically independent and identically distributed in the state ρ . Setting

$$\bar{p}_n = n^{-\frac{1}{2}} (p_1 + p_2 + \dots + p_n) \quad ,$$

$$\bar{q}_n = n^{-\frac{1}{2}} (q_1 + q_2 + \dots + q_n) \quad ,$$

we infer that for every n the distribution operator $\rho_{\bar{p}_n,\bar{q}_n}$ is also the projector onto the span of the function (12). Consequently, the ground state in question being the limit distribution operator for the sequence $\bar{p}_1,\bar{q}_1,\bar{p}_2,\bar{q}_2,\dots$ is stable which completes the proof.

References.

[1] Araki,H., Hamiltonian formalism and the canonical commutation relations in quantum field theory. J.Math.Phys. 1 (1960), p.492--504 .

[2] Bergström,H., On some expansions of stable distribution function, Ark.för mat. 2 (1952), p. 375-378 .

[3] Cushen,C.D. and Hudson,R.L., A quantum-mechanical central limit theorem, J.Applied Math. 8 (1971), p. 454-469.

[4] Dixmier,J., Sur la relation i(PQ-QP)=1 . Comp.Math.13 (1958) p. 263-270 .

[5] Lévy,P., Théorie de l'addition des variables aléatoires, Paris, 1954 .

[6] K.Urbanik, Joint probability distributions of observables in quantum mechanics, Studia Math. 21 (1962), p. 117-133 .

[7] Wigner,E., On the quantum correction for thermodynamic equilibrium, Phys.Rev. 40 (1932), p. 749-759 .

WINTER SCHOOL
ON PROBABILITY
Karpacz 1975
Springer's LNM

472

PREDICTION THEORY IN BANACH SPACES

By A. Weron
Wrocław Technical University

1. Introduction.

The object of this paper is to give the basic formalism and some of the main results in the prediction theory of stochastic processes taking values in an infinite dimensional Banach space. Presented results extend the Wiener-Kolmogorov correlation theory of the best linear least squares prediction for wide sense stationary processes. The concept of stricly stationary processes admitting a prediction was introduced and discussed in [18] but in this paper we will consider only second order processes.

Classical prediction theory is concerned with a stationary sequence, that is, a family X_n of complex-valued random variables on probability space $(\Omega,\bar{B},P)$ that have zero means and finite covariances (X_n,X_k) depending only on $n-k$. One accomplishment of the theory is an analitical characterization of those processes which have the null remote past, that is of the intersection, over all integers, of the closed spans in $L_2(\Omega,\bar{B},P)$ od families $\{X_m, m \leq n\}$. Such processes are called completely indeterministic. A second feature of the classical theory is an analitical characterization of processes which are exactly predictable (deterministic processes).

Remind that a stationary process X_n is completely indeterministic iff there exists a representation

(1.1) $$X_n = \sum_{k=-\infty}^{0} a_k V_{k+n} ,$$

where V_k is an orthogonal sequence, $\Sigma|a_k|^2 < \infty$, or, equivalently, iff its spectral measure m is absolutely continuous with respect to the the Lebesque measure and

(1.2) $$\frac{dm}{dt} = |h(t)|^2 > 0 \quad a.e.,$$

where

$$h(t) = \sum_{k=0}^{\infty} c_k e^{ikt} .$$

A stationary process is deterministic iff its spectral measure m satisfies

(1.3) $$\int_0^{2\pi} \log \frac{dm_a}{dt} dt = -\infty ,$$

where m_a is the absolutely continuous part of m with respect to the Lebesque measure (cf. [19]).

In the extention of prediction theory two important trends are being observed. One trend connected with the set of parameters and second with the space of values of a process. The first will not be considered here and we refer the reader to [5] and [23] for an account of the extrapolation and interpolation theory of stationary processes over locally compact Abelian group, respectively. For the second, we remark that the attempt to extend to infinite-dimensional processes the theory of q-variate processes, developed by Rozanov [14] and Wiener and Masani [25] has atracted the attention of several mathematicians: Gangolli [7], Payen [13], Kallianpur and Mandrekar [9], Nadkarni [12], Chobanyan and Vakhania [2], Schmidt [15] and others.

In one-dimensional theory of stationary processes the family of random variables forms a Hilbert space and consequently, Hilbert space methods play a key role there. Therefore stationary processes are studied as curves in Hilbert space. The idea again occured for non-stationary processes of second order. By generalising the concept of Hilbert space in such a way that the inner product takes values which are no longer scalars (but elements of more general topological space) it is possible to extend the above model to infinite-dimensional stochastic processes. This paper is devoted to the study of Banach space valued stochastic processes of second order (not necessary stationary) as curves in a Loynes space. The space $L(B,H)$ of all linear continuous operators from a Banach space B to a Hilbert space $H = L_2(\Omega,\bar{B},P)$ is a Loynes space and play the same role in our investigations as the $L_2(\Omega,\bar{B},P)$ space in the Wiener-Kolmogorov theory.

In Section 2 we give preliminary results on random elements in Banach spaces. Loynes spaces which first appeared in [10] are considered in Section 3. In the next Section we define Banach space valued stochastic processes, their correlation function and give some examples and applications to operator theory. Section 5 is devoted to the discussions of the general prediction problem and Wold decomposition. In the last Section we show how the classical results (1.1) - (1.3) may be extended to stationary processes with values in Banach spaces.

2. Random elements in Banach spaces.

Let $(\Omega,\bar{B},P)$ be a probability space and E be a complex Banach space with the topological dual E^*. A function $x:\Omega \to E$ is said to be a random element (random E-variable) if it is weakly measurable (for each $f \in E^*$ $f(x(\omega))$ is measurable). The mathematical expectation of random elements is defined by Pettis integral. In this paper we consider only random elements of second order i.e., $f(x(\omega)) \in L_2(\Omega,\bar{B},P)$ for each $f \in E^*$. With each random E-variable we may relate in a natural way the linear transformation of the dual space E^* into the space $L_2(\Omega,\bar{B},P)$:

$$(2.1) \qquad (X f)(\omega) = f(x(\omega)), \quad f \in E^* .$$

The operator X in (2.1) has a closed graph and according to the closed-graph theorem X is continuous. The following fact on X follows, for example from ([17], Th.2).

(2.2) PROPOSITION. *If E is a separable Banach space then the operator X^* adjoint to the operator X, generated by a random E-variable as in (2.1), maps $L_2(\Omega,\bar{B},P)$ into $E \subset E^{**}$.*

From this result we obtain

(2.3) COROLLARY. <u>Let</u> E <u>be a separable Banach space and</u> $x(\omega)$ <u>be a random</u> E-<u>variable</u>. <u>Then its mathematical expectation exists</u>.

<u>Proof</u>. Since constant function $g(\omega) = 1$ belongs to $L_2(\Omega,\bar{B},P)$ thus according to (2.2) $X^*1 \in E$. If we denote $m = X^*1$ then for $f \in E^*$ we obtain

$$f(m) = (X^*1)(f) = (Xf,1) = \int_\Omega (Xf)(\omega)P(d\omega) = \int_\Omega f(x(\omega))P(d\omega) .$$

Consequently, the mathematical expectation of $x(\omega)$ exists and is equal to m.

In view of (2.3) we loose no generality in assuming that mathematical expectations of all random E-variables exist and are equal to zero. The covariance operator of a random element is defined as follows

$$(R\,f)(g) = \int_\Omega \overline{f(x(\omega))}\, g(x(\omega))\, P(d\omega) , \quad f,g \in E^*, \tag{2.4}$$

where $R : E^* \to E^{**}$. It is known ([17], Th.2) that if E is a seperable real Banach space than a linear operator $R : E^* \to E^{**}$ is the covariance operator of a random element iff $(Rf)(f) \geqslant 0$ and $R : E^* \to E$. We refer the reader to Vakhania ([20] and [21]) for an account of the covariance theory of random elements in Banach spaces.

(2.5) DEFINITION. Each linear continuous operator $X : E^* \to L_2(\Omega,\bar{B},P)$ is said to be a <u>generalized random element</u> on E.

However, we note that there exists generalized random elements which are not generated by any random E-variable. For example, the identical operator $I : L_2(0,1) \to L_2(0,1)$ is not generated by any random element in $L_2(0,1)$. If E is a Hilbert space and X is a Hilbert-Schmidt operator than it is generated by a random element. The question when X is generated by random E-variable is discussed in detail in [4].

The mathematical expectation and the covariance operator of a generalized random element are defined in an analogous way as for random elements. From the lemma on factorization (cf. [20], p.135 or [4], (2.4)) the following characterization is obtained.

(2.6) PROPOSITION. <u>An antilinear continuous operator</u> $R: E^* \to E^{**}$ <u>is the covariance operator of a generalized random element iff for</u> $f \in E^*$ $(Rf)(f) \geq 0$.

If X,Y are generalized random elements then we may define their correlation operator by the following relations which is similar to (2.4)

$$(R_{XY}f)(g) = \int_\Omega \overline{(Xf)(\omega)}\ (Yg)(\omega)\ P(d\omega)\ , \quad f,g \in E^{**} .$$

It is easy to see that

$$R_{XY} = Y^* X : E^* \to E^{**} . \tag{2.7}$$

3. <u>Loynes spaces</u>.

Our intention is to consider spaces on which a vector-valued inner product can be defined.

Let $\mathbb{Z}$ be a complex Hausdorff topological vector space satisfying the following conditions:

(3.1) $\mathbb{Z}$ has an involution: i.e. a mapping $z \to z^+$ of $\mathbb{Z}$ to itself with the properties $(z^+)^+ = z$, $(az_1 + bz_2)^+ = \bar{a}z_1^+ + \bar{b}z_2^+$ for all complex a,b ;

(3.2) there is a closed convex cone $\mathbb{P}$ in $\mathbb{Z}$ such that $\mathbb{P} \cap -\mathbb{P} = 0$; then we define a partial ordering in $\mathbb{Z}$ by writing $z_1 \geq z_2$ if $z_1 - z_2 \in \mathbb{P}$;

(3.3) the topology in $\mathbb{Z}$ is compatible with the partial ordering, in the sense that there exists a basic set $\{N_0\}$ of convex neighbourhoods of the origin such that if $x \in N_0$ and $x \geq y \geq 0$ then $y \in N_0$;

(3.4) if $x \in \mathbb{P}$ then $x = x^+$;

(3.5) $\mathbb{Z}$ is complete as a locally convex space ;

(3.6) if $x_1 \geq x_2 \geq \ldots \geq 0$ then the sequence x_n is convergent.

It is clear that these conditions are satisfied by the complex numbers and by the space of all $q \times q$ matrices with the usual topologies. In ([3], Th.1) it is proved that the space $\overline{L}(B,B^*)$ of all antilinear continuous operators from a Banach space into its dual with the weak operator topology also satisfy these conditions.

Now suppose that $\mathfrak{H}$ is a complex linear space. A vector inner product on $\mathfrak{H}$ is a map $x,y \to [x,y]$ from $\mathfrak{H} \times \mathfrak{H}$ into an <u>admissible</u> space $\mathfrak{Z}$ (i.e. satisfying (3.1) - (3.6)) with the following properties:

(3.7) $[x,y] \geq 0$ and $[x,x] = 0$ implies $x = 0$;

(3.8) $[x,y] = [y,x]^{+}$;

(3.9) $[ax_1 + bx_2, y] = a[x_1,y] + b[x_2,y]$ for complex a and b.

When a vector inner product is defined on the space $\mathfrak{H}$ there is a natural way in which $\mathfrak{H}$ may be made into a locally convex topological vector space: a basic set of neighbourhoods of the origin $\{U_0\}$ is defined by

$$\{U_0\} = \{x : [x,x] \in N_0\} . \tag{3.10}$$

(3.11) DEFINITION. The space $\mathfrak{H}$ which is complete in the topology defined by (3.10) with the admissible space $\mathfrak{Z}$ satisfing (3.1) - (3.6), will be called a <u>Loynes space</u>.

These spaces was introduced and investigated by R.M.Loynes in [10]. A complex Hilbert space and the space of all $p \times q$ matrices with the usual topologies are simple examples of Loynes spaces. In ([3], Th.2) we show that the space $L(B,H)$ of all linear continuous operators from a Banach space into a Hilbert space with the strong operator topology is a Loynes space if we define A as an operator adjoint to A ; $(Ax)(y) = \overline{(A\ y)(x)}$, $\mathfrak{P}$ as the set of all non-negative operators : $(Ax)(x) \geq 0$ and $[A,C] = C\ A \in \mathfrak{L}(B,B^{*})$.

It is already clear how a generalized second-order stochastic processes may be defined : X_t is a curve in a Loynes space; such a process is stationary if its correlation function $[X_t, X_s]$ depends only on $(t-s)$. In our recent paper [3] we studied this idea. We have obtained spectral representations and an ergodic theorem for generalized stationary sequences. However, in view of difference between ours and the Hilbert space situation (there are no theorems guaranteeing the existence of the projection onto a closed subspace) we proposed, for linear prediction problems, a model of stationary processes in the special Loynes space $L(B,H)$. This model will be considered in detail in the next section. This situation is in contrast with another generalization of second-order stationarity given in [11], where stochastic processes with values in a Loynes space are studied.

4. B - valued processes.

Let us observe that the set of all generalized random elements may be considered as the linear space $L(B,H)$ if we denote $E^{*} = B$ and $L_2(\Omega,\mathcal{B},P) = H$. In view of the last section this space is a Loynes space with the inner product defined by $[X,Y] = Y^{*}X$. Therefore from (2.7) it follows that the inner product of two generalized random elements X and Y is equal to their correlation operator R_{XY}, which takes values in the admissible space $\mathcal{L}(B,B^{*})$.

(4.1) DEFINITION. By a stochastic process of second order with values in a Banach space (or simply by a B-valued process) we mean a family X_t of elements of $L(B,H)$ indexed by a set T. Its correlation function is defined by the relation $K(t,s) = [X_t,X_s]$. If T is an Abelian group then a B-valued process is called stationary if the function of two variables $K(t,s)$ depends only on $(t-s)$. In the case when T will be a topological group we will assume that $K(t)$ is weakly continuous i.e., for each $b \in B$ the scalar valued function $(K(t)b)(b)$ is continuous.

It is easy to see that B-valued process is stationary iff for each $b \in B$ the one-dimensional process $(X_t b)$ is stationary. We note that B-valued processes appear, in a natural way, as a result of the differentiation of second order fields.

(4.2) Example. Let B be a Banach space and $H = L_2(\Omega,\mathcal{B},P)$. A mapping $\xi : B \to H$ which satisfies the condition $\int_{\Omega} \xi_b P(d\omega) = 0$ for each $b \in B$ is called (one-dimensional) random field of second order over the Banach space B. If the field ξ_b is differentiable in the Fréchet sense at every point $b \in B$ i.e., if there exists a linear operator $(\delta\xi)_b \in L(B,H)$ such that

$$\xi_{b+h} - \xi_b = (\delta\xi)_b\, h + O(\|h\|) \quad \text{as} \quad h \to 0 ,$$

then the derivative $\delta\xi$ is a B-valued process in our sense, defined on $T = B$. Random fields over Banach spaces and their derivatives were studied in [8]. If a field ξ is stationary then for differentiability of ξ at every point $b \in B$ suffices the differentiability of ξ at the point $b = 0$. Let ξ be a stationary field over B satisfying the condition $\int_{B^*} \|b^*\|^2 M(db^*) < \infty$. Then there exists derivative $\delta\xi$ of the field ξ and it is a B-valued stationary process over B (cf. [8], p.307).

(4.3) Example. Let B be a Banach space and V be a linear continuous operator from B to $H = L_2(\Omega,\tilde{B},P)$. If we put

$$X_g = U_g V, \qquad g \in G,$$

where $(U_g)_{g\in G}$ is an unitary representation of an Abelian group G in H then

$$[X_g,X_h] = (U_hV)^* (U_gV) = V^* U_{g-h}V$$

and X_g is a B-valued stationary process. This example is more general because each B-valued stationary process is of this form. Indeed, let H_X denote the closed linear subspace of H spanned over the elements $X_g b$; $g \in G$, $b \in B$. The stationary B-valued process X_g defines in H_X a unitary representation of G. The suitable unitary operators are defined by the formula:

$$U_h X_g b = X_{g+h} b, \qquad g,h \in G, \ b \in B,$$

and for the remaining points of H_X the operators U_h are defined by a natural extension. Consequently $U_g X_e = X_g$.

If we assume that G is a locally compact Abelian LCA group then by the generalized theorem of Stone we obtain

$$X_g = \int_\Gamma \gamma(g)\, E(d\gamma)\, X_e, \tag{4.4}$$

where $E(\cdot)$ is a regular normed and orthogonal spectral family of projectors in H_X defined on the Borel σ-algebra of dual group Γ and $\gamma(g)$ denotes the value of a character $\gamma \in \Gamma$ at $g \in G$. (cf. [4], Section 4). We remark that spectral representations similar to (4.4) may be obtained also for non-stationary processes as in the next example.

(4.5) Example. The B-valued oscilatory processes with continuous time are defined in [24] as

$$Z_t = \int_{-\infty}^{+\infty} e^{itu} a_t(u) \Phi(du) ,$$

where Φ is an orthogonal $L(B,H)$ - valued measure and for each $t \in R$ $a_t(u) \in L_{2,F}$, where $F(A) = [\Phi(A), \Phi(A)]$ for a Borel subset A .

Let us define

$$X_t = \int_{-\infty}^{+\infty} e^{itu} \Phi(du) .$$

Then X_t is a B-valued stationary process and in view of (4.4) $\Phi(A) = E(A)X_o$, where E is the spectral family of projectors associated with the shift group defined by X_t . Let for each t ,

$$Q_t = \int_{-\infty}^{+\infty} a_t(u) E(du) .$$

Hence $Q_t \in L(H,H)$ and by the definition of spectral integrals and the fact $U_sE(A) = E(A)U_s$ we obtain that for each t $Q_tU_sh = U_sQ_th$ for h from the domain of Q_t and for all s . Furthermore

$$Q_tX_t = (\int_{-\infty}^{+\infty} a_t(u)E(du))(\int_{-\infty}^{+\infty} e^{itu}E(du)X_o) = \int_{-\infty}^{+\infty} a_t(u)e^{itu}E(du)X_o = Z_t$$

Hence the B-valued oscilatory process Z_t is obtained by a linear deformation of the B-valued stationary process X_t such that Q_t commutes with the shift group U_t associated with X_t . This fact characterizes B-valued oscilatory processes (cf. [24], Th.3).

According to (4.1) the correlation function of a B-valued process takes values in the admissible space $L(B,B^*)$ of the Loynes space $L(B,H)$. For $L(B,B^*)$ - valued function the notion of positive -definite functions may be defined in several, but equivalent, ways (cf.[22]).

(4.6) THEOREM. <u>An</u> $L(B,B^*)$ - <u>valued function</u> $K(t,s)$ <u>defined on</u> $T \times T$ <u>is the correlation function of a</u> B-<u>valued process iff it is positive-definite i.e., if</u>

$$\sum_{i,j=1}^{N} (K(t_i,t_j)(b_j) \geq 0$$

<u>for each</u> N, $t_1,t_2,\ldots,t_N \in T$ <u>and</u> $b_1,b_2,\ldots,b_N \in B$.

<u>Proof.</u> It is clear that the correlation function is positive - definite. Conversely, let us consider the scalar-valued function $R(g,h)$ of two variables $g = (x,t)$ and $h = (y,s)$ from $B \times T$, given by the following relation

$$R(g,h) = (K(t,s)x)(y) .$$

Since $K(t.s)$ is an operator-valued positive-definite function then $R(g,h)$ is positive-definite, Therefore, in view of the well known fact, there is a Gaussian stochastic process ξ_g in H such that

$$R(g,h) = (\xi_g, \xi_h) .$$

Thus

$$(4.7) \qquad (K(t,s)x)(y) = (\xi(x,t), \xi(y,s)) ,$$

where $\xi(x,t)$ is a H-valued function of two variables $x \in B$ and $t \in T$. By (4.7) the function $\xi(x,t)$ is linear on the first variable if the second is fixed. Hence there exists a linear operator $Y_t : B \to H$, $t \in T$, such that

$$Y_t x = \xi(x,t) , \quad x \in B .$$

Since

$$\|Y_t x\|^2 = (\xi(x,t),\xi(x,t)) = (K(t,t)x)(x) \leq \|K(t,t)\| \|x\|^2$$

thus Y_t is bounded. Consequently Y_t is B-valued process with the correlation function $K(t,s)$.

This theorem has numerous interesting consequences, a few of which will now be presented.

(4.8) THEOREM. *If G is a group, B a complex Banach space and $f : G \to L(B,B^*)$ is positive-definite then there is a Hilbert space H, an operator $A \in L(B,B)$ and a unitary representation U of G on H such that $f(g) = A^* U_g A$.*

Proof. If f is a positive-definite $L(B,B^*)$ - valued function then by (4.6) f is the correlation function of a B-valued process X_t on G. Since f depends only on one variable, X_t is stationary. Consequently by (4.3) we have

$$f(g) = [X_g, X_e] = X_e^* X_g = X_e^* U_g X_e .$$

If we put $X_e = A$ then the theorem is proved.

Theorem (4.8) implies the following Naimark's theorem ([6],p.68) on unitary dilation of an operator-valued positive-definite function. It suffice to put in (4.8) $B = K$ - a Hilbert space and A - an isometric mapping from K on H.

(4.9) COROLLARY. *If G is a group, K is a Hilbert space and $f : G \to L(K,K)$ is positive-definite with $f(e) = I$, there is a Hilbert space $H \supset K$ and an unitary representation U of G on H* such that $f = PU|_K$.

From (4.6) and the generalized theorem of Bochner for operator-valued functions (cf. [22]) we obtain the following characterization of the correlation function of a B-valued stationary process over LCA groups.

(4.10) THEOREM . *If $K(g)$ is a weakly continuous $\overline{L}(B,B^*)$ - valued function on an LCA group G then the following are equivalent:*

(a) $K(g)$ is the correlation function of a B-valued stationary process over G,

(b) $K(g)$ is positive-definite,

(c) there is a unique positive $L(B,B^*)$ - valued regular Borel measure F on the dual group Γ such that for all $g \in G$

$$K(g) = \int_\Gamma \gamma(g)\, F(d\gamma) ,$$

<u>where the integral is meant in the weak sense.</u>

In view of the equivalence of the conditions (a) and (c) of the above result we have the following theorem on factorization of an operator-valued measure (cf. also [1]).

(4.11) THEOREM. <u>Let</u> F <u>be a positive</u> $L(B,B^*)$ - <u>valued measure on the Borel</u> σ-<u>algebra of an LCA group. Then there exist a Hilbert space</u> H , <u>a spectral measure</u> E (<u>i.e. a projection-valued positive measure in</u> $L(H,H)$) <u>and</u> $A \in L(B,H)$ <u>such that for each Borel set</u> D $F(D) = A^* E(D) A$.

<u>Proof</u>. Let F be a positive $L(B,B^*)$ - valued measure on the Borel σ-algebra of an LCA group Γ and $K(g)$ its Fourier transform on the dual group G :

$$K(g) = \int_\Gamma \gamma(g)\, F(d\gamma) .$$

By (4.10) $K(g)$ is the correlation function of a B-valued stationary process over G.Hence by (4.4)

$$K(g) = [X_g, X_e] = [\int_\Gamma \gamma(g) E(d\gamma) X_e, X_e] = \int_\Gamma \gamma(g) X_e^* E(d\gamma) X_e .$$

It follows from the uniqueness in (4.10) (c) that for each Borel set D

$$F(D) = X_e^*\, E(D)\, X_e .$$

Since $X_e \in L(B,H)$ it suffices to put $A = X_e$.

This theorem shows that in Hilbert space case a positive operator-valued measure (semispectral measure) can always be dilated to a spectral measure. More exactly we have the following famous Naimark's theorem on dilation (cf. [6], p.74).

(4.12) COROLLARY. <u>Let</u> K <u>be a Hilbert space and</u> F <u>be a semispectral measure (positive</u> $L(K,K)$ - <u>valued measure and</u> $F(\Gamma) = I$)

on the Borel σ-algebra of an LCA group. Then there is a Hilbert space $H \supset K$ and a spectral measure E in H such that $F = PE|_K$.

5. The prediction problem.

If X_t is a B-valued process we denote by $\underline{M}_X(S)$ a closed linear subspace of $L(B,H)$ spanned by the elements $X_t A$, $t \in S \subset T$ and $A \in L(B,B)$. The space $\underline{M}_X = \underline{M}_X(T)$ is the operator-time-domain of the process. Similarly let $H_X(S)$ denote a closed linear subspace of H spanned by $X_t b$, $t \in S \subset T$ and $b \in B$. The space $H_X = H_X(T)$ is called the vector-time-domain of the process.

In the classical prediction theory the best possible prediction can be expressed with the aid of orthogonal projection. As we noticed in Section 3, for a Loynes space there is no theorem guaranteeing the existance of the projection . But for the special Loynes space $L(B,H)$ the difficulties connected with the projection do not arise because for the linear prediction problem it suffices to define the orthogonal projection on the closed right ideal $\underline{M}_X(S)$ in $L(B,H)$ and by Yood's theorem this projection may be well defined (cf. [3]). More exactly, from Yood's result it follows that for each closed right ideal $\underline{N}$ in $L(B,H)$ there exists a closed subspace $N \subset H$ such that the ideal $\underline{N}$ contains only operators $Z \in L(B,H)$ the range of which is included in N :

$$\underline{N} = \left\{ Z \in L(B,H) \; : \; Zb \in N \quad \text{for} \quad b \in B \right\} .$$

Consequently, there exists a close relation between closed right ideals in $L(B,H)$ and closed subspaces in H . Hence the orthogonal projection onto a closed right ideal $\underline{N}$ is defined as follows:

$$\underline{P}(Z) = P\,Z \quad ,$$

where $Z \in L(B,H)$ and P is the orthogonal projection onto N in the Hilbert space H .

It is clear that the closed right ideal $\underline{M}_X(S)$ in $L(B,H)$ is by Yood's theorem connected with the closed subspace $H_X(S)$ in H . If P_S denotes the orthogonal projection onto $H_X(S)$ then the following definition arises.

(5.1) DEFINITION. The <u>best linear prediction</u> of a B-valued process X_t at a point $u \in T$, based on a subset $S \subset T$ is the element

$$X^{o}_{u,S} = P_S X_u$$

from $L(B,H)$. The operator $Q_{u,S}(X)$ from $L(B,B^*)$, defined by the following relation

$$Q_{u,S}(X) = [X_u - X^{o}_{u,S}, X_u - X^{o}_{u,S}]$$

is the corresponding error involved in this prediction.

From (5.1) follows that for each $b \in B$ and $Z \in \underline{M}_X(S)$

$$\|X_u b - X^{o}_{u,S} b\| \leq \|X_u b - Zb\| .$$

Moreover,

$$X^{o}_{u,S} = \lim_{n\to\infty} \sum_{k=1}^{n} X_{t_k} A_k ,$$

where the limit is meant in the strong operator topology in $L(B,H)$, $t_k \in S$ and $A_k \in L(B,B)$.

(5.2) DEFINITION. Let J be a family of subsets of T. We shall say that B-valued process is J-<u>singular</u> if $\bigcap_{S\in J} \underline{M}_X(S) = \underline{M}_X$ and is J-<u>regular</u> if $\bigcap_{S\in J} \underline{M}_X(S) = (0)$.

Let $L_X = \bigcap_{S\in J} H_X(S)$. It is easy to see that a B-valued process is J-singular iff $L_X = H_X$ and is J-regular iff $L_X = (0)$.

(5.3) <u>Remark</u>. In the classical problem of extrapolation the singularity and regularity of the process are determind by its behaviour on the family J_∞ of intervals $(-\infty,s]$. In the problem of interpolation the families J_C of complements of all compact subsets and J_0 of complements of all singletons are used (cf. [23]). For the problem of extrapolation over groups the following two families arise. If T is a discrete Abelian group and Ψ an arbitrary non-trivial real-valued homomorphism having the image $\Psi(T)$ non-dense in the reals, then J_Ψ denotes the family of all translation over T of the proper subsemigroup $\{t \in T : \Psi(t) \leq 0$, cf. [4] $\}$. If $T = G^+ \times G^-$, where G^+ is a linear ordered group and G^- is a group, then J_+ denotes the family of all sets $\{(g^+, g^-) : g^+ \leq z, g^- \in G^-\}$ (cf. [15], [16]).

The properties of B-valued processes are characterized in the terms of their correlation function. The next result shows that J-regularity and J-singularity may be also characterized in this way.

(5.4) PROPOSITION. <u>Let</u> X_t <u>and</u> Y_t <u>be</u> B-<u>valued processes having the common correlation function</u> $K(t,s)$. <u>Then</u> X_t <u>is</u> J-<u>regular</u> (J-<u>singular</u>) <u>iff</u> Y_t <u>is</u> J-<u>regular</u> (J-<u>singular</u>) <u>and the prediction error operators are equal</u>.

<u>Proof</u>. If B-valued processes have the common correlation function then there exists a unitary operator $U : H_X \to H_Y$ such that $Y_t = UX_t$ for all $t \in T$ (cf. [4], Th.8.3). Since for each $S \in J$, $UH_X(S) = H_Y(S)$,

$$\bigcap_{S\in J} H_Y(S) = \bigcap_{S\in J} U\, H_X(S) = U \bigcap_{S\in J} H_X(S)$$

and consequently X_t is J-regular (J-singular) iff Y_t is J-regular (J-singular). For the proof of the second part of the proposition we remind that if U is an isometry in a Hilbert space H then $U P_N x = P_{UN} Ux$, where P_N is the orthogonal projection on N and $x \in H$. Hence we have

$$Q_{t,S}(Y) = [Y_t - Y^0_{t,S} , Y_t - Y^0_{t,S}]$$

$$= [UX_t - UP_S X_t , UX_t - UP_S X_t] = Q_{t,S}(X) .$$

Let X_t, Y_t be a B-valued processes. Y_t is said to be J-<u>subordinate</u> to X_t if

(a) $\underline{M}_Y(S) \subset \underline{M}_X(S)$ for all $S \in J$,

(b) $\underline{M}_Y \subset \underline{M}_X$.

for stationary B-valued processes we assume else

(c) X_t and Y_t are mutually stationary correlated i.e., $[X_t, Y_s]$ depends only on $t-s$.

Now we get the Wold decomposition theorem. We only sketch the proof, details may be found in [4] and [24].

(5.5) THEOREM. *Let* J *be a family of non-empty subsets of* T *such that for each* $t \in T$ *there is* $S \in J$ *that* $t \in S$. *Let* X_t *be a* B-*valued process. Then there exists a unique decomposition of* X_t *with respect to* J *in the form:*

$$X_t = V_t + W_t ,$$

where

(a) V_t *and* W_t *are mutually orthogonal* B-*valued processes* *i.e.* $[V_t, W_s] = (0)$ *for each* $t,s \in T$,

(b) V_t *and* W_t *are* J-*subordinate to* X_t ,

(c) V_t *is* J-*singular and* W_t *is* J-*regular*.

It X_t *is stationary and the family* J *is closed under translations over the group* T , *then the processes* V_t *and* W_t *are stationary also.*

Proof. Let for each $t \in T$

$$X_t = PX_t + (I-P)X_t = V_t + W_t , \tag{5.6}$$

where P denotes the orthogonal projection onto $L_X = \bigcap_{S \in J} H_X(S)$. Thus V_t and W_t are mutually orthogonal B-valued processes J-subordinate to X_t . From (5.6) we obtain that

$$L_V = H_V(S) = H_V(T) = L_X \qquad \text{for } S \in J .$$

Hence V_t is J-singular. Since $L_W \subset L_X = L_V$ and L_W is orthogonal to L_V therefore W_t is J-regular.

With a B-valued process X_t we may associate a new one by the following linear deformations :

$$Z_t = Q_t X_t \quad \swarrow \quad X_t \quad \searrow \quad Y_t = X_t P_t$$

where $Q_t \in L(H,H)$ and $P_t \in L(B,B)$ for each $t \in T$. The B-valued process Z_t is called _left-associated_ (Y_t is called _right-associated_) with X_t. According to (4.5) B-valued oscilatory processes are left-associated with B-valued stationary processes. If X_t is B-valued process then $Y_t = (X_t b)$, where b is a fixed element of B, is one-dimensional right-associated process with X_t. We conclude this Section with the result on a connection between J-regularity and J-singularity of a B-valued process X_t and processes associated with them.

(5.7) PROPOSITION. _Let_ X_t _be a_ B-_valued process_. _Then_
(a) _if for each_ $b \in B$ _the right-associated process_ $X_t b$ _is_ J-_singular then_ X_t _is_ J-_singular_,
(b) _if_ X_t _is_ J-_regular and for each_ $t \in S$ _and_ $S \in J$ $Q_t(H_X(S)) \subset H_X(S)$ _then the left-associated process_ Z_t _is_ J-_regular_.

Proof. Let each process $X_t b$ be J-singular and suppose that the B-valued process X_t is not J-singular. Then there exists a set $S_1 \in J$ such that $H_X(S_1) \neq H_X(T)$. Hence we may find $t_1 \in T$ and $b_1 \in B$ that $X_{t_1} b_1 \notin H_X(S_1)$. But it contradicts the assumption that all one-dimensional processes are J-singular. Consequently (a) holds. Now if $Z_t = Q_t X_t$ and $Q_t(H_X(S)) \subset H_X(S)$ for $t \in S$ then $H_Z(S) \subset$ $\subset H_X(S)$ for $S \in J$ and therefore $L_Z \subset L_X$. Since $L_X = (0)$, $L_Z = (0)$ and Z_t is J-regular.

6. _Linear extrapolation._

In the last Section we consider B-valued stationary processes with discrete time: X_n, $n = 0, \pm 1, \pm 2, \dots$, Based on ([16], Th.3.3.1) and ([4], Th.10.2) we get the following theorem which generalizes classical results (1.1) and (1.2).

(6.1) THEOREM. _Let_ X_n _be a_ B-_valued stationary process_. _Then the following are equivalent_:
(a) X_n _is completely indeterministic_ (J_∞-_regular_, _see_ (5.3)),
(b) _there exists a representation_

$$X_n = \sum_{k=-\infty}^{n} Z_k A_{n-k} ,$$

where Z_k is an orthogonal stationary process with values in a Hilbert space K i.e., $[Z_k, Z_n] = \delta_{n,k} I_K$, $A_k \in L(B,K)$ for $k = 0,1,2,\ldots$ and

$$\sum_{k=0}^{\infty} \|A_k b\|^2 < +\infty$$

for each $b \in B$,

(c) the spectral measure F (see (4.10)) is absolutely continuous with respect to the Lebesque measure and there exists a Hilbert space K and a sequence of operators $A_k \in L(B,K)$ such that the series

$$h_b(t) = \sum_{k=0}^{\infty} e^{-ikt} A_k b$$

converges for each $b \in B$ in $L_2^K[0,2\pi)$ and for each $b \in B$ we have

$$\frac{d(Fb)(b)}{dt} = \|h_b(t)\|^2 .$$

Proof. (a) $\Longrightarrow$ (b). Let us denote $H_X(-\infty,t] = H_X(t)$. It is well known that

$$H_X(0) = \bigcap_{k=0}^{\infty} H_X(-k) \oplus \left(\bigoplus_{k=0}^{\infty} N_k \right),$$

where N_k $H_X(-k) \ominus H_X(-k-1)$. In view of $J_{\oplus}$- regularity

$$\bigcap_{k=0}^{\infty} H_X(-k) = (0)$$

and consequently

$$P_X(0) = \sum_{k=0}^{\infty} Q_X(k) , \tag{6.2}$$

where $P_X(0)$ and $Q_X(k)$ are the orthogonal projections on $H_X(0)$ and N_k, respectively. Let K be a Hilbert space such that $\dim K = \dim \overline{Q_X(0) H_X}$ and Z_0 an isometry $K \to H_X$ for which

$Z_0 K = \overline{Q_X(0)\,H_X}$. If we put $Z_n = U_X^n Z_0$, where U_X^n is the shift group defined by X_n in H_X then $[Z_k, Z_n] = \delta_{n,k} I_K$ and Z_n is a K-valued stationary process. Moreower $H_Z = H_X$, $H_Z(0) = H_X(0)$ and $U_X^n = U_Z^n$. Let us denote by $A_k = Z_0^* X_k$, then

$$Q_X(k)\, X_0 = Z_{-k} A_k \ .$$

From (6.2) it follows

$$\sum_{k=0}^{\infty} Z_{-k} A_k = \sum_{k=0}^{\infty} Q_X(k) X_0 = P_X(0) \wedge_0 = X_0 \ .$$

Thus

$$X_n = U_X^n X_0 = \sum_{k=0}^{\infty} U_Z^n Z_{-k} A_k = \sum_{k=-\infty}^{n} Z_k A_{n-k} \ .$$

(b) ==> (c) . Let us denote

$$h_b(t) = \frac{1}{\sqrt{2\pi}} \sum_{k=0}^{\infty} e^{-ikt} A_k b \ , \tag{6.3}$$

where A_k are as in the representation (b) of X_n , $t \in [0,2\pi)$ and $b \in B$. It is easy to see that for each $b \in B$ the series in (6.3) is convergent in $L_2^K [0,2\pi)$. The Fourier transform of the function $\|h_b(t)\|^2$ has the following form

$$\int_0^{2\pi} e^{itz} \|h_b(t)\|^2 \, dt = \sum_{n=0}^{\infty} (A_n^* A_{z+n} b)(b) \ . \tag{6.4}$$

On the other hand in view of the condition (b) we have for the correlation function of X_n

$$(K(z)b)(b) = \sum_{n=0}^{\infty} (A_n^* A_{z+n}\, b)(b) \ . \tag{6.5}$$

From (6.4) and (6.5) we obtain that

$$(K(z)b)(b) = \int_0^{2\pi} e^{itz} \|h_b(t)\|^2 \, dt \; .$$

Hence by the uniqueness of the Fourier transform and by (4.10) we conclude that the spectral measure F of the B-valued stationary process X_n is absolutely continuous with respect to the Lebesque measure and that for each $b \in B$

$$\frac{d(Fb)(b)}{dt} = \|h_b(t)\|^2 \; .$$

For the proof of the remainding implication (c) $\Longrightarrow$ (a) we refer the reader to ([4], the proof of Th. 10.2).

Now, as a simple application of the (5.7) (a), we obtain that the following infinite dimensional analogue of the condition (1.3) is sufficient for J_∞-singularity. But as the example given in [4] shows this condition is not necessary.

(6.6) PROPOSITION. *Let* X_n *be a* B-*valued stationary process. If for each* $b \in B$

$$\int_0^{2\pi} \log \left[d(F_a\, b)(b)/du \right] du = -\infty \; ,$$

where F_a *is the absolutely continuous part of the spectral measure, then the* B-*valued process* X_n *is deterministic.*

Finally from (6.1) we may obtain the formula for the linear extrapolation of completely indeterministic B-valued processes.

(6.7) PROPOSITION. *If* X_n *is a completely indeterministic* B-*valued stationary process then the best linear extrapolation and the error are as follows*

$$X^0_{n,(-\infty,0]} = \sum_{k=-\infty}^{0} Z_k A_{n-k}$$

and

$$Q_{n,(-\infty,0]} = \sum_{k=1}^{n} A^{*}_{n-k} A_{n-k} .$$

References.

[1] S.A.Chobanyan, On some properties of positive operator valued measures in Banach spaces, (in Russian), Bull. Acad. Sci. Georg. SSR 57(1970), 273-276.

[2] S.A.Chobanyan and N.N.Vakhania, Wide-sense valued stationary processes in Banach space, (in Russian), Bull. Acad. Sci. Georg. SSR 57(1970), 545-548.

[3] S.A.Chobanyan and A.Weron, Stationary processes in pseudo-hilbertian space, (in Russian), Bull. Acad. Polon. Sci., Sér. Sci. Math. Astronom. Phys. 21(1973), 847-854.

[4] S.A.Chobanyan and A.Weron, Banach space valued stationary processes and their linear prediction, Dissertationes Math. 125(1975).

[5] D.M.Eaves, Prediction theory over discrete Abelian groups, Trans. Amer. Math. Soc., 136(1969), 125-137.

[6] P.A.Fillmore, Notes on operator theory, Van Nostrand Reinh. Math. Studies vol. 30, 1970.

[7] R.Gangolli, Wide-sense stationary sequences of distributions on Hilbert space and the factorization of operator-valued functions, J. Math. Mech. 12(1963), 893-910.

[8] R.Jajte, On derivatives of random fields over a Banach space, Bull. Acad. Polon. Sci., Sér. Sci. Math. Astronom. Phys. 22(1974), 305-311.

[9] G.Kallinapur and V.Mandrekar, Spectral theory of H-valued processes, J. Mult. Anal. 1(1971), 1-16.

[10] R,M.Loynes, Linear operators in VH-spaces, Trans. Amer. Math. Soc. 116(1965), 167-180.

[11] R.M.Loynes, On generalization of second-order stationarity, Proc. Lond. Math. Soc. 15(1965), 385-398.

[12] M,G.Nadkarni, Prediction theory of infinite variate weakly stationary stochastic processes, Sakhya, Series A 32(1970), 145-172.

[13] R.Payen, Functions aléatoires de second ordre a valuers dans un espace de Hilbert, Ann. Inst. H. Poincaré 3(1967), 323-396.

[14] Yu.A.Rozanov, Spectral theory of multi-dimensional stationary processes with discrete time, (in Russian), Usp. Mat. Nauk 13 (1958), 93-142.

[15] F.Schmidt, Spectraldarstellund und Extrapolation einer Klasse von stationären stochastischen Prozessen, Math. Nachr. 47(1970), 101-119.

[16] F.Schmidt, Verallgemeinerte stationäre stochastische Prozesse auf Gruppen der Form $Z \times G^{-}$, Math. Nachr. 57(1973), 327-357.

[17] V.I.Tarieladze, A characterization of a class of covariance operators, (in Russian), Bull. Acad. Sci. Georg. SSR 72(1973),529-532.

[18] K.Urbanik, Prediction of stricly stationary sequences, Coll.Math. 12(1964), 115-129.

[19] K.Urbanik, Lectures on prediction theory, Lecture Notes in Math. Vol. 44, Springer-Verlag 1967.

[20] N.N.Vakhania, Probability distributions on linear space, (in Russian), Mecnieraba , Tbilisi 1971.

[21] N.N.Vakhania, On some question of the theory of probability measures on Banach spaces, Lecture Note Nagoya University, 1973.

[22] A.Weron, On positive-definite operator-valued functions in Banach spaces, (in Russian), Bull. Acad. Sci. Georg. SSR 71(1973), 297-300.

[23] A.Weron, On characterizations of interpolable and minimal stationary processes, Studia Math. 49(1974), 165-183.

[24] A.Weron, Stochastic processes of second order with values in Banach spaces, Trans. of 1974 European Meting of Statisticians and 7-th Prague Conference, (in print).

[25] N.Wiener and P.Masani, The prediction theory of multivariate stochastic processes I, Acta Math. 98(1957), 111-150; II Acta Math. 99(1958), 93-137.

WINTER SCHOOL
ON PROBABILITY
Karpacz 1975
Springer's LNM

472

GEOMETRY AND MARTINGALES IN BANACH SPACES

By W.A.Woyczyński
Wrocław University
and
the Institute of Mathematics, Polish Academy of Sciences

CONTENTS

0. Introduction

This is a survey article on interrelations existing between the metric geometry of Banach spaces and the theory of martingales with values in Banach spaces.

The first chapter deals with the geometric notion of dentability, and gives the exposition of the results, originally due to Rieffel, Maynard, Chatterji and others. Despite the recent result of Huff and Morris (cf. Remark 1.1) which might suggest to some that the notion of dentability belongs to the history, and is bound to fall into oblivion, we feel that what is written below shows transparently that it is very natural a notion in the context of martingales, even though its appearance might have seemed somewhat mysterious in the context of the Radon-Nikodym Property.

The second chapter deals with two classical notions of metric geometry in Banach spaces, namely uniform smoothness, and uniform convexity. Here the works of Pisier and Assouad have shown that some of the theorems proved earlier by Woyczyński for sums of independent random vectors in such Banach spaces carry over to a more general situation of martingales providing even a characterization of those geometric properties in the language of martingales.

Chapter 3 is about certain geometric conditions on Banach spaces that are invariants of isomorphisms, that is do not change under equivalent renorming. Finite tree property and super-reflexivity, the notions introduced by James, turned out to be those properties that are most intimately related to the martingale theory as it was displayed by recent results of Kwapien and Pisier which are dealt with in this chapter.

At last, Chapter 4 takes care of Banach lattices which are dentable and surveys some of the results of Szulga and Woyczyński on submartingales in such spaces.

We took pains to provide the proofs of all the theorems that appear below, and that provide an illustration for the interplay between geometry and martingales. In contrast, in general, we skipped the more difficult proofs of results which are purely geometric or purely probabilistic in nature. We tried (not too hard) to improve on the existing proofs whenever we found it feasible. As this survey was compiled for educational purposes, for the benefit of students we included into references some entries that are not explicitly quoted in the text, just to make the list more complete.

U s a g e : R stands for real numbers, N for positive integers and we use asterisks identifying formulas locally. $w \equiv \omega$.

Taking an opportunity, we would like to thank all the friends who helped me while this work was being prepared and especially Patrice Assouad, Jo Diestel, Tadeusz Figiel and Gilles Pisier who supplied us with the manuscripts of yet unpublished papers of theirs.

1. Dentability, Radon-Nikodym Theorem and Martingale Convergence Theorem

Rieffel [1] introduced the following geometric concept of a dentable subset in a Banach space $\mathfrak{X}$ ($\mathfrak{X},\mathfrak{Y}$ will be always Banach spaces throughout the paper, separable if one considers measurable functions with values in $\mathfrak{X},\mathfrak{Y}$).

DEFINITION 1.1 . $A \subset \mathfrak{X}$ is <u>dentable</u> if for every $\varepsilon > 0$ there exists an $x \in A$ such that $x \notin \overline{\text{conv}}\,(A - B_\varepsilon(x))$.

Here $\overline{\text{conv}}$ stands for a closed convex hull and $B_\varepsilon(x) = \{x \in \mathfrak{X} : \|x\| \leq 1\}$.

DEFINITION 1.2 . The Banach space $\mathfrak{X}$ is said to be <u>dentable</u> if every bounded subset of $\mathfrak{X}$ is dentable.

It is evident that dentability is invariant under equivalent renorming, and that, as the matter of fact, in order to check that the space $\mathfrak{X}$ is dentable it is sufficient to check that all closed convex subsets of $\mathfrak{X}$ are dentable. Indeed, we have the following.

PROPOSITION 1.1. (Rieffel [1]). <u>If</u> $\overline{\text{conv}}\,A^{-}$, $A \subset \mathfrak{X}$, <u>is dentable then</u> A <u>is dentable</u>.

<u>Proof</u>. Let $\varepsilon > 0$ be given. Then one can find $y \in \overline{\text{conv}}\,(A)$ such that

$$y \notin Q \overset{\text{df}}{=} \overline{\text{conv}}\,[\overline{\text{conv}}\,(A) - B_{\varepsilon/2}(y)] \,.$$

Now let $x \in A - Q$. Such an x does exist because $y \in \overline{\text{conv}}\,A - Q$, Q is closed, convex so that Q can not contain A . Evidently $x \notin \overline{\text{conv}}\,(A - B_\varepsilon(x))$ because $x \in B_{\varepsilon/2}(y)$ and $(A - B_\varepsilon(x)) \subset Q$.

<u>Remark</u> 1.1. Actually Davis and Phelps [1] have shown that $\mathfrak{X}$ is dentable if and only if after arbitrary equivalent renorming the new unit ball is dentable (cf. the proof of Proposition 1.1.), and recently Huff and Morris [1] have proved that $\mathfrak{X}$ is dentable if and only if every bounded closed set in $\mathfrak{X}$ has an extreme point (in its convexification). It is an open question whether the dentability is equivalent to the Krein-Milman Property (i.e. every bounded, closed, convex subset of $\mathfrak{X}$ possesses an extreme point).

It is easy to see that the unit balls of Banach spaces c_0 ,

$L_1[0,1]$, $C[0,1]$ are not dentable. On the other hand, a simplest example of a dentable set is a set A which has fixed denting point x i.e. such that for any $\varepsilon > 0$, $x \notin \overline{\text{conv}}\,(A - B_\varepsilon(x))$. It is not very hard to see that the <u>strongly exposed point</u> of A (i.e. such that $x^* \in X^*$ and $\alpha \in R$ such that $\{y: x^* y = \alpha\} \cap A = \{x\}$ and if for $(y_n) \subset A$, $x^* y_n \to \alpha$ then $\|y_n - x\| \to 0$) is always the denting point of A and that every denting point of A is the extreme point of $\overline{\text{conv}}\,A$. As the matter of fact one can also show (Davis [1], p. 0.9), that the above example is general enough i.e. that each closed, convex, bounded and non-empty set A in a dentable space has always a denting point.

Another characterization of dentable sets A uses slices of $A \subset X$ i.e. sets of the form $(x^* x \geq \beta) \cap A$ with non-empty $(x^* x > \beta) \cap$

PROPOSITION 1.2. (cf. Davis [1]). <u>A closed, bounded, convex and non-empty $A \subset X$ is dentable if, and only if for any $\varepsilon > 0$ there is a slice of A with diameter less than ε .</u>

<u>Proof</u>. If A is dentable then for arbitrary $\varepsilon > 0$ and we can find $x \in A$ such that $x \notin \text{conv}\,(A - B_\varepsilon(x))$ and so $x^* \in X^*$, $\alpha \in R$, in a way that

$$x^* x > \alpha > \sup\left\{x^* x : x \in \overline{\text{conv}}\,(A - B_\varepsilon(x))\right\} .$$

In particular

$$(x^* x \geq \alpha) \cap A \subset B_\varepsilon(x) \quad ,$$

what ends the proof in one direction. The other implication is straightforward.

The proposition given below facilitates checking the dentability of many sets and spaces. In particular, it implies that all weakly compactly generated Banach spaces (thus all separable duals) are dentable. (Dentability of all reflexive Banach spaces follows from Theorem 1.1. and older work on Radon-Nikodym Theorem done by Dunford, Morse, Pettis and Phillips).

PROPOSITION 1.3. (Asplund and Namioka [1]). <u>Weakly compact convex sets in a separable Banach space are dentable.</u>

<u>Proof</u>. It is sufficient to show that for any $\varepsilon > 0$ and arbitrary $K \subset \mathfrak{X}$, weakly compact and convex, one can find a closed convex, one can find a closed convex $C \subset K$ with diameter of $K - C$ less than ε.

Denote by P the weak closure of the set of extreme points in K, and let

$$P \subset \bigcup_{i=1}^{\infty} B_{\varepsilon/4}(x_i) \quad , \qquad x_i \in P .$$

Because P is weakly compact, hence second category in itself, there is $x \in P$ and a weakly open neighbourhood W of x such that

$$P \cap B_{\varepsilon/4}(x) \supset W \cap P \neq \emptyset \ .$$

Denote

$$K_1 = \overline{\text{conv}}\ (P - W) \quad , \qquad K_2 = \overline{\text{conv}}\ (W \cap P) .$$

Evidently, by Krein-Milman theorem, K is the convex hull of weakly compact sets K_1, K_2 and, furthermore, $K_1 \neq K$ because the extreme points of K_1 lie within $P - W$ (cf. Kelley and Namioka 15.2). Now, define

$$C_r = \left\{ tk_1 + (1-t)k_2 : k_i \in K_i \ , \ r \leq t \leq 1 \right\} \ , \qquad 0 \leq r \leq 1.$$

Clearly, the C_r are weakly compact, convex, and increase as $r \to 0+$. $C_0 = K$ and $C_1 = K_1$. Finally, note that $C_r \neq K$ for all $0 < r \leq 1$, for if $C_r = K$, then each extreme point z of K has the form

$$z = \lambda x_1 + (1-\lambda)x_2 \quad , \qquad x_i \in K_i \ , \ \lambda \in [r,1] .$$

Hence, $z = x_1 \in K_1$, contradicting $K_1 \neq K$. Notice that, if $y \in K - C_r$ then y is of the form $y = \lambda x_1 + (1-\lambda)\, x_2$, $x_i \in K_i$, $\lambda \in [0,r]$ and

$$\|y - x_2\| = \|\lambda x_1 + (1-\lambda)x_2 - x_2\| = |\lambda|\,\|x_1 - x_2\| \leq r\|x_1 - x_2\| ,$$

so each $y \in K - C_r$ lies within distance $r.\text{diam}\ (K_2)$ of K_2. But diam $(K_2) \leq \varepsilon/2$, thus as $r \to 0t$, C_r has the desired property diam $(K - C_r) < \varepsilon$.

Maynard [1] introduced the apparently weaker notion of the σ-dentable set.

DEFINITION 1.3. $A \subset \mathfrak{X}$ is said to be σ-dentable if for each $\varepsilon > 0$ there exists $x \in A$ such that $x \notin \sigma\text{-conv}\,(A - B_\varepsilon(x))$.

Here

$$\sigma\text{-conv}\,(B) \overset{\mathrm{df}}{=} \left\{ \sum_{i=1}^{\infty} \lambda_i b_i : \lambda_i \geq 0 \, , \ \Sigma\, \lambda_i = 1 \, , \ b_i \in B \right\} , \ B \subset \mathfrak{X}$$

Actually, we have the following result due to Davis and Phelps [1].

PROPOSITION 1.4. *The Banach space* $\mathfrak{X}$ *is dentable if, and only if it is* σ-*dentable*.

Proof. Obviously, it is sufficient to show that if $\mathfrak{X}$ is σ-dentable then it is also dentable. Assume to the contrary that there exists in $\mathfrak{X}$ the set A which is bounded and not dentable. Take $x \in \mathfrak{X}$ so that $(x+A) \cap (-x-A) = \emptyset$. Then $B = \text{conv}\,(x+A\ ,\ -x-A)$ is a closed, convex, symmetric and also non-dentable by Proposition 1.1. so that we may assert that the unit ball $B_1(0) \subset \mathfrak{X}$ is not dentable. Indeed, were it dentable, the closed convex body $\overline{B + B_1(0)} \subset \mathfrak{X}$ which generates on $\mathfrak{X}$ the norm equivalent to the original would also be dentable and by Proposition 1.1. $B + B_1(0)$ would also be dentable, but it is not (because B is not dentable one can find $\varepsilon > 0$ such that for all $x \in B$ $x \in \overline{\text{conv}}\,(B - B_\varepsilon(x))$ so that if $x+y \in B+B_1(0)$ then also

$$x + y \in \overline{\text{conv}}((y+B) - B_\varepsilon(y+x)) \subset \overline{\text{conv}}((B_1(0)+B) - B_\varepsilon(y+x)) \ \ ).$$

Now, we shall show that non-dentability of $B_1(0)$ implies non-σ-dentability of int $B_1(0)$ what, in turn, would contradict the σ-dentability of $\mathfrak{X}$.

Take $\varepsilon > 0$ such that for each $x \in B_1(0)$, $x \in \overline{\text{conv}}(B_1(0)-B_\varepsilon(x)$ If $\|x\| < 1 - \varepsilon/4$ then for some $\lambda > 0$, $\|\lambda x\| < 1$, $\|x-\lambda x\| > \varepsilon/4$, $\|x+\lambda x\| > \varepsilon/4$. Thus, $x \in \text{conv}(B_1(0) - B_{\varepsilon/4}(x))$. If $1 > \|x\| > 1 - \varepsilon/4$

then $B_{\varepsilon/4}(x) \subset B_\varepsilon(x/\|x\|)$ so that $x/\|x\| \in \overline{\text{conv}}(B_1(0) - B_{\varepsilon/4}(x))$. For small ε, 0 is an interior point of $\overline{\text{conv}}(B_1(0) - B_{\varepsilon/4}(x))$, so that entire segment $[0, x/\|x\|)$ is in the interior of that set. In particular

$$x \in \text{conv}\ (\text{int}\ B_1(0) - B_{\varepsilon/4}(x))$$

so that $\text{Int}\ B_1(0)$ is not σ-dentable.

The equivalence proved in the above proposition shows that it is sufficient to know what separable Banach spaces are dentable in order to know what are all dentable Banach spaces. In particular $\mathfrak{X}$ is dentable if and only if each closed separable subspace of $\mathfrak{X}$ is dentable. More precisely we have the following .

PROPOSITION 1.5. (Diestel [1]). <u>Dentability is the separably determined property i.e.</u> $A \subset \mathfrak{X}$ <u>is</u> σ-<u>dentable if, and only if each countable subset of</u> A <u>is</u> σ-<u>dentable</u>.

<u>Proof</u>. Assume that A is not σ-dentable. Then we can find an $\varepsilon > 0$ such that for each $x \in A$, $x \in \sigma\text{-conv}\ (A - B_\varepsilon(x))$. Thus, letting $x^{(1)} \in A$ be arbitrary, one gets $(x_n^{(1)}) \subset A$ such that $x^{(1)} = \Sigma\, \lambda_n\, x_n^{(1)}$ and $\|x^{(1)} - x_n^{(1)}\| \geq \varepsilon$. Now, apply to each $x_n^{(1)}$ the same denial of the σ-dentability of A. Reiterating this procedure one gets an infinite tree in A which is countable and, by definition, not σ-dentable.

Now, we pass to the investigation of interrelations between dentability and validity of certain measure-theoretic and martingale-theoretic theorems in Banach spaces.

DEFINITION 1.4. We say that the Banach space $\mathfrak{X}$ has the Radon-Nikodym Property (RNP) if every $\mathfrak{X}$-valued measure m on (Ω, Σ) for which the total variation measure

$$|m|(E) \overset{\text{df}}{=} \sup \left\{\Sigma\ \|m(E_i)\| \ :\ E_i \in \Sigma \text{ disjoint},\ \ \cup E_i = E\right\},$$

$E \in \Sigma$, is finite, and which is absolutely continuous with respect to a finite positive measure μ, admits with respect to μ a Bochner integrable density.

DEFINITION 1.5. Let (Ω,Σ,P) be a probability space and let $\Sigma_1 \subset \Sigma_2 \subset \ldots \subset \Sigma$ be an increasing sequence of sub-σ- algebras of Σ. A sequence of strongly measurable $\mathfrak{X}$ - valued functions (M_n), M_n mes Σ_n, is said to be a $[(\Sigma_n)]$ martingale if $E(M_{n+1}/\Sigma_n) = M_n$, $n = 1,2,\ldots$.

DEFINITION 1.6. We say that the L_p - Martingale Convergence Theorem holds in $\mathfrak{X}$ $(\mathfrak{X} \in (MCT_p))$, $1 \leq p < \infty$, if for each $\mathfrak{X}$ - valued martingale (M_n, Σ_n) such that $\sup_n E\|M_n\|^p < \infty$, there is an $M_\infty \in L^p(\Omega,\Sigma,P ; \mathfrak{X})$ such that $M_n \to M_\infty$ a.s. in norm.

Remark 1.2. By pure martingale theory (cf. Chatterji [3]) one can prove (and we omit the proof following our principles from the introduction) that if $\mathfrak{X} \in (MCT_{p_0})$ for some $1 \leq p_0 < \infty$ then $\mathfrak{X} \in (MCT_p)$ for all p from that interval. Moreover, if $1 < p < \infty$ then $M_n \to M_\infty$ also in $L^p(\Omega,\Sigma,P ; \mathfrak{X})$. We shall utilize these facts freely in the sequel.

Now, we deem the following two examples illuminating:

Example 1.1. (Chatterji [1]). $\mathfrak{X} = L^1(0,1) \notin (MCT_1)$. Indeed, let Σ_n be the binary Borel algebra in $(0,1)$ generated by intervals $(m/2^n, (m+1)/2^n)$, $0 \leq m \leq 2^n-1$, $n = 1,2,\ldots$.
Define

$$M_n(w) = 2^n (\chi_{(0,(m+1)/2^n]} - \chi_{(0,m/2^n]})$$

if $w \in (m/2^n, (m+1)/2^n)$ and 0 elsewhere. It is easy to check that $(M_n, \Sigma_n, n \geq 1)$ is a martingale with values in $L^1(0,1)$ and moreover

$$\|M_n(w)\| \equiv 1 \quad \text{a.e.} \ ,$$

$$E(\|M_n(w)\|^p) = 1 \ , \quad n \geq 1 \ ,$$

$$E(\sup_n \|M_n(w) - M_{n-1}(w)\|) = 1 \ , \quad M_0 = 0 \ .$$

But if $w \neq p/2^q$ then $X_n(w)$ does not go to any limit either weakly or strongly.

<u>Example</u> 1.2. (Lewis [1]). $\mathfrak{X} = c_0 \notin$ (RNP) . Indeed, let (Ω,Σ,μ) be a finite positive measure space containing no atoms. We construct the measure $m : \Sigma \to c_0$ (the elements of c_0 will be denoted $(a_{n,i})$, $n = 1,2,\ldots$, $2^n \leq i < 2^{n+1}$) in the following manner:

$$m : \Sigma \ni E \to m(E) = (\mu(E \cap E_{n,i})) \in c_0$$

where $(E_{n,i})$, $n = 1,2,\ldots$, $2^n \leq i < 2^{n+1}$ is a sequence of measurable sets such that $\mu(E_{n,i}) = 2^{-n}\,\mu(\Omega)$, and such that $E_{n,i}$ is the disjoint union of $E_{n+1,2i}$ and $E_{n+1,2i+1}$. Such a sequence $(E_{n,i})$ does exist in view of non-atomicity of μ .

Evidently $\|m(E)\| \leq \mu(E)$ so that m is absolutely continuous with respect to μ and has finite total variation.

However, m has no Bochner integrable density with respect to μ . In fact, if it had, say $f : \Omega \to c_0$, then writing $(e_{n,i})$ for the standard basis in l_1 , for each $i = 2,3,\ldots$ and $E \in \Sigma$, we would have that

$$\int_E f(w)\; e_{n,i}\; \mu(dw) = \mu(E \cap E_{n,i})$$

so that, there would exist, for each i , a μ - null set $C_i \subset \Omega$ such that $f(w)e_{n,i} = \chi_{E_{n,i}}(w)$ for $w \notin C_i$. Choose $w_0 \in \Omega - \cup_i C_i$. By the very construction of $(E_{n,i})$, $\chi_{E_{n,i}}(w) = 1$ for infinitely many indices i , so $\lim_i f(w)e_{n,i} \neq 0$, what gives the desired contradiction.

The following is the main theorem of this chapter. The equivalence of the first and second conditions follows from the work of Rieffel [1], Maynard [1] and Davis & Phelps [1]. The short and direct proof of this equivalence was recently found by Huff [1]. The equivalence of the third and second condition was proved by Chatterji [3]. We prove the theorem in the circular fashion giving the new proofs for second and third implications.

THEOREM 1.1. <u>For a Banach space</u> $\mathfrak{X}$ <u>the following properties are equivalent</u>

(D) $\mathfrak{X}$ <u>is dentable</u> ,

(RNP) $\mathfrak{X}$ <u>has the Randon-Nikodym Property</u>

(MCT_p) <u>the</u> L_p - <u>Martingale Convergence Theorem holds in</u> $\mathfrak{X}$.

Proof. We shall prove the following implications:
(D) $\Longrightarrow$ (RNP) $\Longrightarrow$ (MCT_p) $\Longrightarrow$ (D) .

(D) $\Longrightarrow$ (RNP)

Assume that $\mathfrak{X}$ is dentable, and m is an $\mathfrak{X}$ - valued measure on (Ω,Σ) with finite total variation, and which is absolutely continuous with respect to a finite positive measure μ (the proof is the same even if μ is σ-finite). The task is to find a Bochner Σ - measurable $f : \Omega \to \mathfrak{X}$ such that for each $E \in \Sigma$

$$m(E)= \int_E f(w)\ \mu(dw) .$$

First observation m has with respect to μ locally almost dentable average range i.e. for each $E \in \Sigma$ and an arbitrary $\varepsilon > 0$ there exists $F \subset E$ such that $\mu(E - F) < \varepsilon$ and that

$$AR(F) \overset{df}{=} \left\{\frac{m(F')}{\mu(F')} : F' \subset F\ ,\ \mu(F') > 0\right\} \subset \mathfrak{X}$$

is dentable. Indeed, because of dentability of $\mathfrak{X}$ it is sufficient to prove that $AR(F)$ is bounded, and the last statement may be verified as follows:

The total variation $|m|$ is a finite positive measure on Σ whic is also absolutely continuous with respect to μ , and, so, by the "real-valued" Radon-Nikodym Theorem, for some $\varphi \in L^1(\Omega,\Sigma,\mu\ ;\ R)$

$$|m|\ (E) = \int_E \varphi(w)\ \mu(dw) \quad , \qquad E \in \overline{fe} ,$$

so that given $\varepsilon > 0$ there is a constant K such that $\varphi(w) < K$ on $E_0 \in \Sigma$, $\mu(\Omega - E_0) < \varepsilon$. Thus, given $E \in \Sigma$ and $\varepsilon > 0$ we take $F = E \cap E_0$. Then $\mu(E - F) \leq \mu(\Omega - E_0) < \varepsilon$ and for each $F' \subset F = E \cap E_0$ with $\mu(F') > 0$ we'll have

$$\left\| \frac{m(F')}{\mu(F')} \right\| \leq \frac{|m|(F')}{\mu(F')} \leq K .$$

Second observation. For any $\varepsilon > 0$ and $E \in \Sigma$, $\mu(E) > 0$, one can find $F \subset E$, $\mu(F) > 0$, and an $x \in \mathfrak{X}$ such that

$$\|m(F') - x\mu(F')\| < \varepsilon\mu(F')$$

for all $F' \subset F$.

Indeed, by First observation, there is an $E_d \subset E$ of positive measure μ such that $AR(E_d)$ is dentable, and, in particular, for an ε given above one can find

$$x = \frac{m(F_o)}{\mu(F_o)} \in AR(E_d) \quad , \qquad F_o \subset E_d \, , \quad \mu(F_o) > 0 \, ,$$

such that

$$x \notin Q \overset{df}{=} \overline{\text{conv}}\,(AR\,(E_d) - B_\varepsilon(x)) \, .$$

Now, either F_o can be taken as F or not. If it can, we are trough, and if not we can find $E_1 \subset F_o$ such that $\mu(E_1) \geq 1/k_1$ (and let k_1 be the smallest integer for which such E_1 exists) and

$$\frac{m(E_1)}{\mu(E_1)} \in Q$$

Now, either $F_1 = F_o - E_1$ is the looked for F or not. If yes, we are trough, if not, we find $E_2 \subset F_1$ as above. So, either we find our F in a finite number of steps, or, by induction, we choose a sequence (E_i) of pair-wise disjoint subsets of F_o and a sequence of (minimal in the above sense) integers $k_i \uparrow \infty$ (because $\mu(F_o)<\infty$) such that

$$\mu(E_i) \geq \frac{1}{k_i} \qquad \text{and} \qquad \frac{m(E_i)}{\mu(E_i)} \in Q \, ,$$

and such that if

$$E' \subset F_o - \bigcup_{i=1}^{n} E_i \qquad \text{and} \qquad \frac{m(E')}{\mu(E')} \in Q$$

then $\mu(E') < 1/(k_n-1)$ (minimality of k_n!). But in this case we can surely take

$$F = F_0 - \bigcup_{i=1}^{\infty} E_i \subset F_0 - \bigcup_{i=1}^{n} E_i \ , \qquad n = 1,2,\ldots ,$$

because:

If for any $F' \subset F$, $\mu(F') > 0$, $m(F')/\mu(F') \in Q$, then we would have that for each $n = 1,2,\ldots$, $\mu(F') < 1/(k_n-1)$ what would yield $\mu(F') = 0$, a contradiction.

If F were of μ measure 0 then also $m(F) = 0$ and by convexity argument we would have that

$$x = \frac{m(F_0)}{\mu(F_0)} = \frac{m(\bigcup E_i)}{\mu(\bigcup E_i)} = \sum_{i=1}^{\infty} \frac{m(E_i)}{\mu(E_i)} \ \frac{\mu(E_i)}{\mu(\bigcup E_i)} \in Q$$

again a contradiction.

Third observation. For each $\varepsilon > 0$ one can find a sequence $(x_i) \subset X$ and an, at most countable, partition $(E_i) \subset \Sigma$ of Ω such that

(*) whenever $F \subset E_i$, $\mu(F) > 0$ then

$$\|m(F) - x_i\mu(F)\| \leqslant \varepsilon\mu(F) \ .$$

Indeed, using repeatedly the second Observation, either we exhaust Ω in the finite number of steps or else we find a sequence (E_i) of pairwise disjoint subsets of Ω , a sequence $(x_i) \subset X$, and a non-decreasing sequence $k_i \uparrow \infty$ of (minimal as in the Second Observation) integers such that for each $i = 1,2,\ldots$ (*) holds, $\mu(E_i) \geqslant 1/k_i$, and if for some n

$$E \subset \Omega - \bigcup_{i=1}^{n} E_i$$

is such that for some $x \in X$ and all $F \subset E$, $\mu(F) > 0$

$$\|m(F) - x\mu(F)\| < \varepsilon\mu(F)$$

then $\mu(E) < 1/(k_n-1)$. But in the latter case $\mu(\Omega - \bigcup E_i) = 0$ because, otherwise, using the Second Observation we could find $E \subset \Omega - \bigcup E_i$, $\mu(E) > 0$ as above, what would mean that for each n $\mu(E) < 1/(k_n-1)$ i.e. $\mu(E) = 0$, a contradiction.

E p i l o g u e . Now, we can complete the proof of the implication (D) ==> (RNP) constructing the density f as follows:

Let Π $(\ni\pi)$ be the directed family of finite partitions of Ω into sets positive measure, and put

$$f_\pi = \sum_{E\in\pi} \frac{m(E)}{\mu(E)} \chi_E .$$

If we menage to show that $\lim_\pi f_\pi$ exists in $L^1(\Omega,\Sigma,\mu ; \mathfrak{X})$ then, clearly $f = \lim_\pi f_\pi$ is the looked for density, for then, im particular, for each $E \in \Sigma$, $\mu(E) > 0$

$$\int_E f\, d\mu = \lim_{\pi\in\Pi} \int_E f_\pi\, d\mu ,$$

and because for $\pi \geq \{E, \Omega-E\}$

$$\int_E f_\pi\, d\mu = m(E)$$

So, the proof will be complete as soon as we show that (f_π) is $L^1(\Omega,\Sigma,\mu ; \mathfrak{X})$ - Cauchy.

Take $\varepsilon > 0$. Because $|m|$ is a finite positive measure, absolutely continuous with respect to μ, we can choose $E \subset \Omega$ such that $|m|(\Omega-E) < \varepsilon/6$ and $\delta > 0$ such that $|m|(F) < \varepsilon/6$ whenever $\mu(F)<\delta$.

In the decomposition $E = E_i$ (for $\varepsilon/6\, \mu(E)$ instead of ε) from the Third Observation take n such that

$$\mu(E - \bigcup_{i=1}^{n} E_i) < \delta$$

and

$$\pi_0 = \left\{E_1,\dots,E_n , E - \bigcup_{i=1}^{n} E_i , \Omega - E\right\} .$$

Then for arbitrary $\pi \geq \pi_0$ which is evidently of the form

$$\pi = \left\{F_{ij} : 1 \leq i \leq n , 1 \leq j \leq k_i\right\} \cup \left\{G_1,\dots,G_l\right\} \cup \left\{F_1,\dots,F_k\right\} ,$$

$$\bigcup_{j=1}^{k_i} F_{ij} = E_i \quad , \qquad\qquad 1 \leq i \leq n \quad ,$$

$$\int \|f_\pi(w) - f_{\pi_0}(w)\| \; \mu(dw) = \sum_{i=1}^{n} \sum_{j=1}^{k_i} \left\| \frac{m(F_{ij})}{\mu(F_{ij})} - \frac{m(E_i)}{\mu(E_i)} \right\| \mu(F_{ij}) +$$

$$+ \sum_{i=1}^{l} \left\| \frac{m(G_i)}{\mu(G_i)} - \frac{m(E - \bigcup_{i=1}^{n} E_i)}{\mu(E - \bigcup_{i=1}^{n} E_i)} \right\| \mu(G_i) + \sum_{i=1}^{k} \left\| \frac{m(F_i)}{\mu(F_i)} - \frac{m(\Omega - E)}{\mu(\Omega - E)} \right\| \mu(F_i) \leq$$

$$\sum_{i=1}^{n} \sum_{j=1}^{k_i} \|m(F_{ij}) - x_i \mu(F_{ij})\| + \sum_{i=1}^{n} \|m(E_i) - x_i \, \mu(E_i)\|$$

$$+ 2 \sum_{i=1}^{l} \|m(G_i)\| + 2 \sum_{i=1}^{k} \|m(F_i)\| \leq \sum_{i=1}^{n} \sum_{j=1}^{k_i} \frac{\varepsilon}{\sigma\mu(E)} \mu(F_{ij})$$

$$+ \sum_{i=1} \frac{\varepsilon}{\sigma\mu(E)} \mu(E_i) + 2 \, |m|(E - \bigcup_{i=1}^{n} E_i) + 2 \, |m|(\Omega - E) \leq$$

$$\frac{\varepsilon}{6} + \frac{\varepsilon}{6} + 2\,\frac{\varepsilon}{6} + 2\,\frac{\varepsilon}{6} = \varepsilon \quad .$$

Q. E. D.

$$(\mathrm{RNP}) \Longrightarrow (\mathrm{MCT}_p)$$

We assume that $\mathfrak{X}$ has the Radon-Nikodym Property, $p > 1$, and that (M_n, Σ_n) is an $\mathfrak{X}$ - valued L^p - bounded martingale on (Ω, Σ, μ) (μ - a probability).

Now, the martingale property yields that the $\mathfrak{X}$ - valued, countably additive measure

$$m_n(E) = \int_E M_n(w) \; \mu(dw) \quad ,$$

$E \in \Sigma_n$, $n = 1,2,\ldots$, extend to a finitely additive set function, say m , on the algebra $\bigcup_{i=1}^{\infty} \Sigma_i$. The total variation of m on any set $E \subset \Sigma_i$, say $E \in \Sigma_n$,

$$|m|(E) = \int_E \|M_n\| \, d\mu \leq [\mu(E)]^{\frac{1}{q}} \sup_n \left(\int_\Omega \|M_n\|^p\right)^{1/p} < \infty$$

($q = p/(p-1) < \infty$) and, moreover, the above inequality shows that $|m|$ is countably additive on $\bigcup \Sigma_n$, thus extends to a measure onto $\Sigma_\infty = \sigma(\bigcup \Sigma_n)$ and is a.c.with r.to μ. So m is also c.a.on Σ_n, extends to Σ_∞ and is absolutely continuous with respect to μ because $\|m(E)\| \leq |m|(E)$, $E \in \Sigma_n$. (For all the properties of vector set functions and their variations used here see Dunford and Schwartz [1], Ch. III and Ch. IV. 10). Thus by the RNP one can find $g \in L^1(\Omega, \Sigma_\infty, \mu \,;\, \mathfrak{X})$ such that

$$m(E) = \int_E g \, d\mu \quad , \qquad E \in \Sigma_\infty \, .$$

Evidently for $n = 1,2,\ldots$, $E(g/\Sigma_n) = M_n$ because for each $E \in \Sigma_n$

$$\int_E g d\mu = m(E) = m_n(E) = \int_E M_n d\mu \quad ,$$

and by general theorem on conditional expectation

$$M_n = E(g/\Sigma_n) \to E(g/\Sigma_\infty) = g$$

a.s. and in $L^p(\Omega, \Sigma, \mu \,;\, \mathfrak{X})$ (cf. Neveu [1], Prop. V-2-6).

$$(MCT_p) \Longrightarrow (D)$$

Assume the space $\mathfrak{X}$ is not dentable i.e. by Proposition 1.4. not σ - dentable. Then there exists a bounded, closed and convex $A \subset \mathfrak{X}$ which is not σ - dentable i.e. such that for some $\varepsilon > 0$ and each $x \in A$ there are positive numbers $\alpha_i(x)$, $\Sigma\, \alpha_i(x) = 1$, and $a_i(x) \in A$, $i = 1,2,\ldots$ such that

$$\inf_i \|x - a_i(x)\| > \varepsilon$$

and

$$x = \Sigma\, \alpha_i(x) \; a_i(x) \quad .$$

We shall construct now the martingale M_n with values in A bounded which diverges a.s thus contradicting (MCT_p). We proceed by induction and construct a sort of tree as in the bicture below

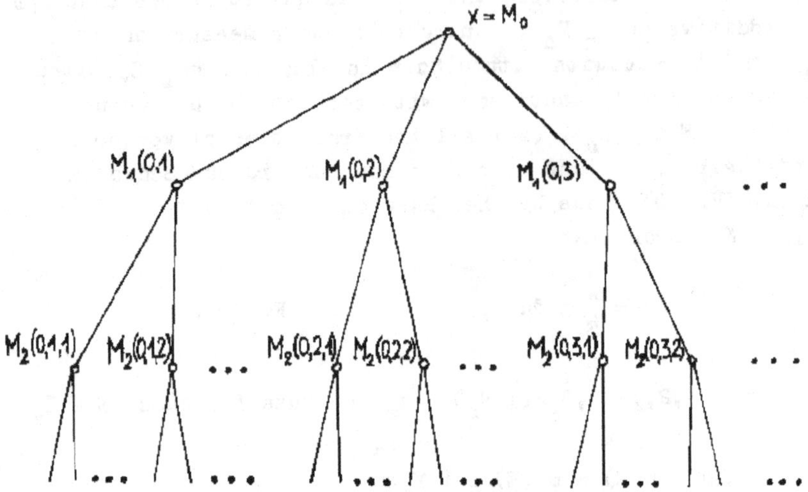

Take (N is the set of positive integers)

$$\Omega = N \times N \times \dots , \quad \Sigma = \sigma(2^N \times 2^N \times \dots)$$

$$P = \underleftarrow{\lim} P_n$$

$$\Sigma_n = \underbrace{2^N \times 2^N \times \dots \times 2^N}_{n\text{ - times}} \times N \times N \times \dots$$

and define

$$M_n(w) = M_n(i_0, i_1, i_2, \dots) = M_n(i_0, \dots, i_n)$$

where M_0 is any $x \in A$ a.s. and, inductively, given M_n and P_n we put

$$M_{n+1}(i_0, i_1, \dots, i_n, i_{n+1}) = a_{i_{n+1}}(M_n(i_0, \dots, i_n))$$

$$P_{n+1}(\{i_0,i_1,\dots,i_{n+1}\}) = P_n(\{i_0,\dots,i_n\})\ \alpha_{i_{n+1}}(M_n(i_0,\dots,i_n)).$$

The very construction causes M_n to be a martingale which has values in the bounded set $A \subset \mathfrak{X}$ and which is evidently divergent as for any $w = (i_0,i_1,\dots) \in \Omega$

$$\|M_{n+1}(w) - M_n(w)\| > \varepsilon \ , \qquad n = 1,2,\dots \ .$$

Remark 1.3. It should be of interest to find a direct proof of the implication (D) $\Longrightarrow$ (MCT) . The first idea is to take a bounded (even Paley-Walsh cf. next section) martingale in $\mathfrak{X}$ which diverges on the set of positive measure, use the uniformization-of-almost-sure-divergence Lemma 1 from Musiał, Ryll-Nardzewski and Woyczyński [1] and then construct from the values of that martingale a sort of minimal non-dentable bounded subset of $\mathfrak{X}$. However, somehow we could not get through with this idea.

2. Uniform convexity and uniform smoothness.

The notion of uniform convexity was introduced by Clarkson [1]

DEFINITION 2.1. Let $\mathfrak{X}$ be a Banach space of dimension at least 2. The modulus of convexity of $\mathfrak{X}$ is defined as

$$\delta_{\mathfrak{X}}(\varepsilon) = \inf\left\{1 - \left\|\frac{x+y}{2}\right\| : \|x\| = \|y\| = 1,\ \|x-y\| = \varepsilon \right\} ,$$

$0 \leq \varepsilon \leq 2$. $\mathfrak{X}$ is said to be uniformly convex if $\delta_{\mathfrak{X}}(\varepsilon) > 0$ for $\varepsilon > 0$. $\mathfrak{X}$ is said to be q - uniformly convex if $\delta_{\mathfrak{X}}(\varepsilon) \geq C\varepsilon^q$, $q \geq 2$, for some constant C .

Example 2.1. L^p - space, $p \geq 1$ is $p \vee 2$ - uniformly convex (cf. Milman [1], that survey article is also recommended as a general reference on metric geometry of Banach spaces, for more information about moduli of convexity and smoothness of Orlicz and other spaces cf. Figiel [1], and Day [2]) .

The notion of uniform smoothness was introduced by Day [1] .

DEFINITION 2.2. Let $\mathfrak{X}$ be a Banach space of dimension at least 2.

The modulus of smoothness of $\mathfrak{X}$ is defined as

$$\rho_{\mathfrak{X}}(\tau) = \sup\left\{\left\|\frac{x+y}{2}\right\| + \left\|\frac{x-y}{2}\right\| - 1 : \|x\| = 1,\ \|y\| = \tau\right\}.$$

$\mathfrak{X}$ is said to be uniformly smooth if $\rho_{\mathfrak{X}}(\tau) = o(\tau)$, $\tau \to 0$. $\mathfrak{X}$ is said to be p - uniformly smooth, $1 < p \leq 2$, if $\rho_{\mathfrak{X}}(\tau) \leq C\tau^p$ for some constant C.

It is evident that the above notions are not invariant under equivalent renorming of the Banach space. So we say that the Banach space $(\mathfrak{X}, \|.\|)$ is uniformly convexifiable [q-uniformly convexifiable, uniformly smoothable, p-uniformly smoothable] if it admits an equivalent to $\|.\|$ norm, that is uniformly convex [q-uniformly convex, uniformly smooth, p-uniformly smooth].

It is the well known fact that $\mathfrak{X}$ is uniformly convex if, and only if the dual space $\mathfrak{X}^*$ is uniformly smooth but Lindenstrauss [1] proved even more

PROPOSITION 2.1. For any Banach space $\mathfrak{X}$ the modulus of smoothness of $\mathfrak{X}$ is the function conjugate in the sense of Young to the modulus of convexity of $\mathfrak{X}$ i.e.

$$\rho_{\mathfrak{X}^*}(\tau) = \sup\left\{\frac{\tau\varepsilon}{2} - \delta_{\mathfrak{X}}(\varepsilon) : 0 \leq \varepsilon \leq 2\right\} \qquad (\tau > 0).$$

Proof. At the beginning let us make an observation that for every positive ε and τ

$$(*) \qquad \delta_{\mathfrak{X}}(\varepsilon) + \rho_{\mathfrak{X}^*}(\tau) \geq \frac{\tau\varepsilon}{2}.$$

Indeed, if $x, y \in \mathfrak{X}$, $\|x\| = \|y\| = 1$, $\|x-y\| = \varepsilon$, and $x^*, y^* \in \mathfrak{X}^*$ are such that $\|x^*\| = \|y^*\| = 1$, and $x^*(x+y) = \|x+y\|$, $y^*(x-y) = \|x-y\|$ then

$$2\rho_{\mathfrak{X}^*}(\tau) \geq \|x^* + \tau y^*\| + \|x^* - \tau y^*\| - 2$$

$$\geq x^*x + \tau y^*x + x^*y - \tau y^*y - 2 = x^*(x+y) + \tau y^*(x-y) - 2 = \|x+y\| + \tau\varepsilon - 2$$

so that

$$2 - \|x+y\| \geq \tau\varepsilon - 2\,\rho_{\mathfrak{X}^*}(\tau) \quad ,$$

and (*) follows.

Now, let $x^*, y^* \in \mathfrak{X}^*$ satisfy the conditions $\|x^*\| = 1$, $\|y^*\| = \tau$, and let $\alpha > 0$. There exist $x, y \in \mathfrak{X}$ such that

$$\|x\| = \|y\| = 1 \quad , \qquad x^*x + y^*x \geq \|x^* + y^*\| - \alpha$$

$$x^*y - y^*y \geq \|x^* - y^*\| - \alpha \quad .$$

Therefore

$$\|x^* + y^*\| + \|x^* - y^*\| \leq x^*x + y^*x + x^*y - y^*y + 2\alpha$$

$$= x^*(x+y) + y^*(x-y) + 2\alpha \leq \|x+y\| + \tau\,\|x-y\| + 2\alpha$$

$$\leq 2 + 2 \sup\,(\varepsilon\tau/2 - \delta_{\mathfrak{X}}(\varepsilon) \;:\; 0 \leq \varepsilon \leq 2) + 2\alpha$$

so that we get our Proposition in view of the arbitrariness of α .

COROLLARY 2.1. _For every Banach space_ $\mathfrak{X}$ _and every_ $\tau \geq 0$ _the modulus of smoothness_

$$\rho_{\mathfrak{X}}(\tau) \geq (1 + \tau^2)^{1/2} - 1 \quad ,$$

the right hand side being the (easily somputable) modulus of smoothness of the Hilbert space (notice, its asymptotics at 0 _is_ τ^2)

COROLLARY 2.2. $\mathfrak{X}$ _is_ q - _uniformly convex if and only if_ $\mathfrak{X}^*$ _is_ p - _uniformly smooth_ $(1/p + 1/q = 1)$.

Example 2.2. The L_p - space, $p > 1$, is $p \wedge 2$ - uniformly smooth.

PROPOSITION 2.2. (Assouad [1], Hoffmann-Jørgensen [1]). _A Banach space_ $\mathfrak{X}$ _is_ p - _uniformly smooth if, and only if there exists a con-_

stant $L > 0$ _such that for all_ $x,y \in \mathfrak{X}$

$$\|x+y\|^p + \|x-y\|^p \leq 2\,\|x\|^p + C\,\|y\|^p \quad ,$$

(i.e. _we are on one side of the parallelogram equality_).

Proof. Let us first make an observation that $\mathfrak{X}$ is p - uniformly smooth if, and only if there exists a constant K such that for all $x,y \in \mathfrak{X}$

$$\left\|\frac{x+y}{2}\right\| + \left\|\frac{x-y}{2}\right\| \leq \|x\|\;(1 + K\,\frac{\|y\|^p}{\|x\|^p}) \;.$$

Now, assume $\mathfrak{X}$ is p - uniformly smooth. Then

$$\tfrac{1}{2}[\,\|x+y\| - (\|x\| + \|y\|) + \|x-y\| - (\|x\| - \|y\|)\,] \leq \|x\|\;K\,\frac{\|y\|^p}{\|x\|^p} \quad .$$

If $\|y\| \leq \|x\|$ then $\|x+y\| \vee \|x-y\| \leq 2\|x\|$, and one gets that

$$\frac{1}{2}\,(\|x+y\|^p + \|x-y\|^p)$$

$$\leq \frac{1}{2}\,[(\|x\| + \|y\|)^p + (\|x\| - \|y\|)^p] + p(2\|x\|)^{p-1}\,\|x\|\;K\,\frac{\|y\|^p}{\|x\|^p}$$

$$\leq \|x\|^p + \|y\|^p + p\,2^{p-1}\,K\,\|y\|^p = \|x\|^p + (1 + p\,2^{p-1}\,K)\,\|y\|^p \quad ,$$

in view of two "real", and easy to check inequalities

$$u^p - v^p \leq pu^{p-1}(u-v) \quad , \qquad u,v \geq 0\;, \quad p \geq 1\;,$$

$$\tfrac{1}{2}(\,|u+v|^p + |u-v|^p) \leq |u|^p + |v|^p \;, \qquad u,v \in R\;, \quad p \in [1,2].$$

On the other hand if $\|y\| \geq \|x\|$ then $\|x+y\| \vee \|x-y\| \leq 2\|y\|$ so that

$$\frac{1}{2}\,(\|x+y\|^p + \|x-y\|^p) \leq \|x\|^p + 2^p\|y\|^p \quad ,$$

so that we get the desired inequality with

$$C = 2^p \vee [1 + p\, 2^{p-1} K]$$

Conversely, if that inequality is fulfilled the $\mathfrak{X}$ is p - uniformly smooth because

$$\rho_{\mathfrak{X}}(\tau) = \sup \left\{ \left\|\frac{x+y}{2}\right\| + \left\|\frac{x-y}{2}\right\| - 1 : \|x\| = 1, \quad \|y\| = \tau \right\}$$

$$\leq \sup \left\{ \left(\left\|\frac{x+y}{2}\right\| + \left\|\frac{x-y}{2}\right\| \right)^p - 1 : \|x\| = 1 , \quad \|y\| = \tau \right\}$$

$$\leq \sup \left\{ \frac{1}{2} \left(\|x+y\|^p + \|x-y\|^p \right) - 1 : \|x\| = 1 , \ \|y\| = \tau \right\} \leq c\tau^p .$$

Remark 2.1. Woyczyński [1], in connection with investigation of random vector series in Banach spaces (cf. also further Remarks), introduced the (G_α) type for a Banach spaces namely $\mathfrak{X} \in (G_\alpha)$, $0 < \alpha \leq 1$, if there exists a mapping

$$G : \left\{ x \in \mathfrak{X} : \|x\| = 1 \right\} \to \left\{ x^* \in \mathfrak{X}^* : \|x^*\| = 1 \right\}$$

such that

$$(G_\alpha^{I}) \qquad \|G(x)\| = \|x\|^\alpha ,$$

$$(G_\alpha^{II}) \qquad G(x)x = \|x\|^{1+\alpha} ,$$

$$(G_\alpha^{III}) \qquad \|G(x) - G(y)\| \leq C\|x-y\|^\alpha ,$$

$x, y \in \mathfrak{X}$. Hoffmann-Jørgensen [1] proved that $\mathfrak{X} \in (G_\alpha)$ if, and only if $\mathfrak{X}$ is $(1+\alpha)$ - uniformly smooth, and gave other alternative descriptions of p - uniform smoothness.

Now, we turn to the investigation of interrelations between uniform smoothness, uniform convexity and martingales.

THEOREM 2.1 (Assouad [1]). *A Banach space* $\mathfrak{X}$ *is* p - *uniformly*

<u>smooth if, and only if there exists a positive constant</u> K <u>such that for any two random vectors in</u> $L^p(\Omega,\Sigma,P\,;\,\mathfrak{X})$: X_1 mes Σ_1 , X_2 mes Σ_2, $\Sigma_1 \subset \Sigma_2 \subset \Sigma$ <u>satisfying the condition</u>

$$(*) \qquad \|2X_1 - E(X_2/\Sigma_1)\| \geq \|X_1\| \quad \text{a.s.}$$

<u>we have the inequality</u>

$$(**) \qquad E(\|X_2\|^p - \|X_1\|^p/\Sigma_1) \leq KE(\|X_2-X_1\|^p/\Sigma_1)$$

<u>Proof</u>. Assume that $\mathfrak{X}$ is p - uniformly smooth. Then the Jensen inequality and (*) yield that

$$E(\|2X_1-X_2\|^p/\Sigma_1) \geq \|2X_1 - E(X_2/\Sigma_1)\|^p \geq \|X_1\|^p \quad ,$$

so that, by Proposition 2.2. putting $x = X_1$, $y = X_2-X_1$, and averaging conditionally on Σ_1, we that

$$E(\|X_2\|^p + \|X_1\|^p/\Sigma_1) \leq E(\|X_2\|^p + \|2X_1-X_2\|^p/\Sigma_1)$$

$$\leq 2E(\|X_1\|^p/\Sigma_1) + CE(\|X_2-X_1\|^p/\Sigma_1)$$

what gives (**).

Now, conversely, assume that (**) holds for any X_1,X_2 as above. So, in particular, given $x,y \in \mathfrak{X}$, we can take $X_1 = x$, $X_2 = x+\varepsilon y$, (ε - a Bernoulli random variable), Σ_1 - trivial. Then (*) is clearly satisfied because $E(X_2/\Sigma_1) = X_1$, and for such X_1,X_2 (**) becomes the inequality

$$E(\|x+\varepsilon y\|^p - \|x\|^p) \leq KE\|\varepsilon y\|^p \quad , \qquad x,y \in \mathfrak{X} \, ,$$

which, in view of Proposition 2.2, means that $\mathfrak{X}$ is p - uniformly smooth.

DEFINITION 2.3. An adapted sequence of random vectors in $\mathfrak{X}$, $\mathfrak{X}_n$ mes Σ_n, $\Sigma_n \subset \Sigma_{n+1} \subset \ldots \subset \Sigma$, $n = 1,2,\ldots$, is said to be the <u>norm supermartingale</u> if for each $n = 1,2,\ldots$

$$\|2X_n - E(X_{n+1}/\Sigma_n)\| \geq \|X_n\|$$

Here are the examples of norm super-martingales.

Example 2.3. Each martingale (M_n,Σ_n) with values in any Banach space is a norm super-martingale.

Example 2.4. If (M_n,Σ_n) is a martingale with values in any Banach space, and (X_n) is a sequence of positive, decreasing and predictable (X_{n+1} mes Σ_n) real random variables, then the sequence (X_nM_n) is a norm super-martingale (an application of this fact will be given later on in this section).

Example 2.5. Any positive real-valued super-martingale is also a norm super-martingale and some "rapidly" growing submartingales are norm super-martingales as well.

COROLLARY 2.3. The Banach space $\mathfrak{X}$ is p - uniformly smooth if, and only if there exists a constant K such that for any $\mathfrak{X}$ - valued norm super-martingale (X_n,Σ_n) , $n = 0,1,\ldots$, in L^p

$$\sup_n E\|X_n\|^p \leq E\|X_0\|^p + K \sum_{n=1}^{\infty} E\|X_{n+1}-X_n\|^p$$

THEOREM 2.2. If $\mathfrak{X}$ is the p - uniformly smooth Banach space, $1 < p \leq 2$, and (M_n) an $\mathfrak{X}$ - valued martingale such that $\Sigma E\|M_{n+1}-M_n\|^p < \infty$, then there exist a random vector $M_\infty \in L^p(\Omega,\Sigma_\infty,P;\mathfrak{X})$ such that $M_n \to M_\infty$ a.s. and in L^p.

Proof. The proof consists of application of Corollary 2.3. Theorem 1.1. and the fact that every p - uniformly smooth space is reflexive and thus dentable (cf. Milman [1], p. 79).

Applying the above theorem and the Kronecker Lemma to the martingale

$$M'_n = \sum_{i=1}^{n} \frac{M_i - M_{i-1}}{i}$$

we get the following

COROLLARY 2.4. (The strong law of large numbers for martingales) *If* (M_n,Σ_n) *is a martingale with values in a* p - *uniformly Banach space such that* $M_0 = 0$ *and*

$$\sum_{i=1}^{\infty} \frac{E\,\|M_i - M_{i-1}\|^p}{i^p} < \infty$$

then

$$\frac{M_n}{n} \to 0$$

almost surely and in L^p .

More theorems on asymptotic behavior of vector valued martingales will be given in another paper by the author.

Remark 2.2. The Theorem 2.2 and Corollary 2.4 generalize Woyczynski's results obtained in [1] , [2] for sums of independent random vectors (cf. Remark 2.1).

It is also worth noticing that, despite the impression that might be given by the above results, the p - uniform smoothness is *not* the property which is uniquely related to general martingales (as opposed to the p-type of the Banach space which is related to martingales with independent increments i.e. sums of independent random vectors. Remind that the Banch space $\mathfrak{X}$ is said to be of type p , $1 < p \leq 2$ (cf. Hoffmann-Jørgensen [2] and Maurey and Pisier [1]) if there exists a positive constant C such that for any $n \in N$ and any independent zero mean random vectors $X_1,\dots,X_n$ with values in $\mathfrak{X}$

$$E\,\|\sum_{i=1}^{n} X_i\|^p \leq C \sum_{i=1}^{n} E\,\|X_i\|^p \quad).$$

Indeed, from the proof of the Theorem 2.1 and Corollary 2.3 we immediately get that *the Banach space* $\mathfrak{X}$ *is* p - *uniformly smooth if, and only if there exists a positive constant* C *such that for any* $n \in N$, *any* $x \in \mathfrak{X}$ *and any independent zero mean random vectors* $X_1,\dots,X_n$ *with values in* $\mathfrak{X}$

$$E \|x + \sum_{i=1}^{n} X_i\|^p \leq \|x\|^p + C \sum_{i=1}^{n} \|X_i\|^p \quad .$$

Despite the formal similarity the notions of p - uniform smoothness and of type p do not coincide. Of course the above discussion shows that every p - uniformly smooth Banach space is of type p. Hovever not conversely (even after renorming). R.James has constructed an example of a Beck convex Banach space (thus of type p for some $p > 1$) which is non-reflexive (so that it can not be p - uniformly smooth for any $p > 1$).

Now, we shall proceed more rapidly in the case of uniformly convex spaces (cf. Assouad [1]).

THEOREM 2.3. *For the Banach space* $\mathfrak{X}$ *the following conditions are equivalent*:

(i) $\mathfrak{X}$ *is* q - *uniformly convex* ;

(ii) *there exists a constant* $L > 0$ *such that for any* $x, y \in \mathfrak{X}$

$$\|x+y\|^q + \|x-y\|^q \geq 2\|x\|^q + L\|y\|^q$$

(i.e. *we are on the other side of the parallelogram equality*);

(iii) *there exists a constant* $K > 0$ *such that for any* $\mathfrak{X}$ - *valued martingale* $(M_n\ ,\ n = 0,1,\dots) \subset L^q$

$$\sup_n E\|M_n\|^q \geq E\|M_0\|^q + K \sum_{n=1}^{\infty} E\|M_{n+1} - M_n\|^q \quad ;$$

(iv) *there exists a constant* $K > 0$ *such that for any* $\mathfrak{X}$ - *valued* r.v.s. X_1 mes Σ_1, X_2 mes Σ_2, $\Sigma_1 \subset \Sigma_2$ (*in* L^q) *such that*

$$\|2X_1 - E(X_2/\Sigma_1)\| = \|X_1\|$$

we have the inequality

$$E\ (\|X_2\|^q - \|X_1\|^q/\Sigma_1) \geq KE\ (\|X_2 - X_1\|^q/\Sigma_1) \quad .$$

Proof: We prove the following implications (i) => (iv) => (iii) => (ii) => (i) .

(i) => (iv).

Assume that $\mathfrak{X}$ is q - uniformly convex. By Theorem 1.3 from Milman [1] it is reflexive, and so is $\mathfrak{X}^*$ which additionally is p - uniformly smooth $(1/p+1/q = 1)$ in view of Corollary 2.2.

Let Y_1^* and Y_2^* be two random vectors in $L^p(\Omega,\Sigma,P;\ \mathfrak{X}^*)$. Then

$$X_1^* = E(Y_1^*/\Sigma_1) \quad , \qquad X_2^* - X_1^* = Y_2^* - E(Y_2^*/\Sigma_1)$$

give rise to a martingale X_1^*, X_2^* with values in $\mathfrak{X}^*$. Now, in view of Theorem 2.1

$$E\ (Y_1^*X_1 + Y_2^*\ (X_2-X_1))$$

$$= E\ [X_1^*X_1 + X_1^*(X_2-X_1) + (X_2^*-X_1^*)X_1 + (Y_2^* - E(Y_2^*/\Sigma_1))(X_2-X_1)]$$

$$= E\ [X_1^*X_1 + X_1^*(X_2-X_1) + (X_2^*-X_1^*)X_1 + (X_2^*-X_1^*)(X_2-X_1)]$$

$$= E\ X_2^*X_2 \leq E\ \|X_2^*\|\ \|X_2\| \leq \frac{1}{p}\ E\ \|X_2^*\|^p + \frac{1}{q}\ E\ \|X_2\|^q$$

$$\leq \frac{1}{p}\ E\ \|X_1^*\|^p + \frac{1}{p}\ K\ E\ \|X_2^*-X_1^*\|^p + \frac{1}{q}\ E\ \|X_2\|^q$$

$$\leq E\ (\frac{1}{p}\ \|Y_1^*\|^p + \frac{1}{p}\ 2^{p+1}\ K\ \|Y_2^*\|^p + \frac{1}{q}\ \|X_2\|^q)\ .$$

Because of arbitrariness of Y_1^* and Y_2^* we get that

$$E\ (\|X_1\|^q + \frac{1}{c}\ \|X_2-X_1\|^q) \leq E\ \|X_2\|^q\ ,$$

the last inequality being motivated by the inequality

$$E\ (Y_1^*\ X_1) \leq E\ \|Y_1^*\|\ \|X_1\| \leq \frac{1}{p}\ E\ \|Y_1^*\|^p + \frac{1}{q}\ E\ \|X_1\|^q\ ,$$

similar inequality for $E(Y_2^*(X_2-X_1))$ and the fact that if for given reals c,b , the inequality $ab \leq a^p/p + c$ holds for all real a then $c \leq b^q/q$.

This ends the proof of this implication.

(iv) => (iii)

Follows by summing up and averaging.

(iii) => (ii)

Follows by putting $M_0 = x$, $M_1 = x+\varepsilon y$, $M_n = M_1$, $n = 2, \dots$, (ε - Bernoulli).

(ii) => (i)

We apply (ii) to the vectors

$$x = \|u\|v + \|v\|u \quad , \qquad y = \|u\|v - \|v\|u \quad ,$$

thus getting that

$$\|2\|u\|v\|^q + \|2\|v\|u\|^q \geq 2\ \|\|u\|v + \|v\|u\|^q + L\|\|u\|v - \|v\|u\|^q$$

wherefrom

$$2^{q+1}\|u\|^q\|v\|^q \geq 2\|\|u\|v + \|v\|u\|^q + L\|\|u\|v - \|v\|u\|^q$$

so that, for any u,v, $\|u\| = \|v\| = 1$, and $\|u-v\| = \varepsilon$ we get that

$$1 - \|\tfrac{u\ v}{2}\|^q \geq \frac{L}{2^{q+1}}\varepsilon^q \quad ,$$

and because of the real inequality $1-\alpha^q \leq q(1-\alpha)$, $0 \leq \alpha \leq 1$, we get that

$$\delta_{\mathfrak{X}}(\varepsilon) = \inf\left\{1 - \|\tfrac{u+v}{2}\| \ : \|u\| = \|v\| = 1, \|u-v\| = \varepsilon\right\}$$

$$\geq \frac{1}{q}\inf\left\{ 1 - \|\tfrac{u+v}{2}\|^q \ : \|u\| = \|v\| = 1, \|u-v\| = \varepsilon\right\} \geq \frac{L}{q2^{q+1}}\varepsilon^q$$

so that $\mathfrak{X}$ is q - uniformly convex.

As for as p - uniformly smoothable and p - uniformly convexifiable Banach spaces are concerned we have the following theorem due to Pisier [1] in the case of Paley-Walsh martingales

THEOREM 2.4. (a) <u>The Banach space</u> $\mathfrak{X}$ <u>is</u> q - <u>uniformly convex-</u>

<u>ifiable if, and only if there exists a constant</u> $C > 0$ <u>such that for any</u> $\mathfrak{X}$ - <u>valued martingale</u> (M_n)

$$(*) \qquad C \sup_n E \|M_n\|^q \geq E \|M_0\|^q + \sum_{n=0}^{\infty} E \|M_{n+1} - M_n\|^q$$

(b) <u>The Banach space</u> $\mathfrak{X}$ <u>is</u> p - <u>uniformly smoothable if, and only if there exists a constant</u> C <u>such that for any</u> $\mathfrak{X}$ - <u>valued martingale</u> (M_n)

$$\sup_n E \|M_n\|^p \leq C \,[E \|M_0\|^p + \sum_{n=0}^{\infty} E \|M_{n+1} - M_n\|^p] \,.$$

<u>Proof</u>. The "only if" parts of (a) and (b) are proved by the straightforward application of Corollary 2.3 and Theorem 2.3. Now, we prove the "if" part of (a) . In the similar manner one gets the proof of the "if" part of (b).

Assume that (*) holds and define the new norm on $\mathfrak{X}$ by means of the formula

$$\Box x \Box = \inf \left(C \sup E \|M_n\|^q - \sum_{n=0}^{\infty} E \|M_{n+1} - M_n\|^q\right)^{1/q}$$

where the infimum is taken over all the Paley-Walsh martingales (M_n) with values in $\mathfrak{X}$, starting at x i.e. such that $M_0 = EM_n = x$, and for which $\sup_n E\|M_n\|^q < \infty$. Remind that the <u>Paley-Walsh martingale</u> is the dyadic martingale on $\Omega = [+1,-1] \times [+1,-1] \ldots$ with natural consecutive σ-fields Σ_n and Bernoulli product probability.

Evidently, $\Box.\Box$ is positively homogeneous, and besides, for all $x \in \mathfrak{X}$

$$\|x\| \leq \Box x \Box \leq C^{\frac{1}{q}} \|x\| \,.$$

The left-hand side inequality follows from (*) , and the right-hand side from the definition of $\Box.\Box$, if one takes $M_n = x$, $n = 0,1,\ldots$.

So, in view of the above inequality, and Theorem 2.3, it is enaugh to show that for all $x,y \in \mathfrak{X}$

$$\Box \frac{x+y}{2} \Box^{q} + \|\frac{x-y}{2}\|^{q} \leq \frac{\Box x \Box^{q} + \Box y \Box^{q}}{2}$$

which will show at the same time the convexity of the set $\{x : \Box x \Box \leq 1\}$ i.e. the triangle inequality for $\Box . \Box$.

Let $x, y, \in \mathfrak{X}$. By the definition of new norm, for any $\alpha > 0$ there exist Paley-Walsh martingales (X_n) and (Y_n) in $\mathfrak{X}$ such that

$$X_0 = x \ , \qquad \sup E \|X_n\|^q < \infty$$

$$Y_0 = y \ , \qquad \sup E \|Y_n\|^q < \infty$$

and

$$C \sup_n E \|X_n\|^q - \sum_{n=0}^{\infty} E \|X_{n+1} - X_n\|^q \leq \Box x \Box^q + \alpha$$

$$C \sup_n E \|Y_n\|^q - \sum_{n=0}^{\infty} E \|Y_{n+1} - Y_n\|^q \leq \Box y \Box^q + \alpha$$

Now, we construct the martingale Paley-Walsh (Z_n) as follows:

$$Z_0 = \frac{x+y}{2}$$

$$Z_n = (\frac{1+\varepsilon_1}{2}) X_{n-1}(\varepsilon_2, \varepsilon_3, \ldots) + (\frac{1-\varepsilon_1}{2}) Y_{n-1}(\varepsilon_2, \varepsilon_3, \ldots) \ .$$

Evidently, $\sup E\|Z_n\|^q < \infty$, and

$$\Box \frac{x+y}{2} \Box^q \leq C \sup_n E\|Z_n\|^q - \sum_{n=0}^{\infty} E\|Z_{n+1} - Z_n\|^q$$

$$\leq C \sup \frac{1}{2}(E\|X_n\|^q + E\|Y_n\|^q) - \|\frac{x-y}{2}\|^q - \sum_{n=0}^{\infty} \frac{1}{2} E\|X_{n+1} - X_n\|^q + E\|Y_{n+1} - Y_n\|^q)$$

$$\leq \frac{\Box x \Box^q + \Box y \Box^q}{2} + \alpha - \|\frac{x-y}{2}\|^q$$

what proves the theorem, in view of arbitrariness of $\alpha > 0$.

A p p l i c a t i o n s . As one of applications of the above theory we would like to show a proof of the following theorem on (non -random) series in p - uniformly smooth Banach spaces. For real series this is due to Steinitz and for L^p to Kadec [1] (and rediscovered later by Drobot in the case of Hilbert spaces). It can also be given a non-probabilistic proof similar to that of Kadec as indicated T. Figiel (oral communication) (cf. also Assouad [1] for arbitrary modulus of smoothness case).

THEOREM 2.5. (Assouad [1]). Let $\mathfrak{X}$ be a p - uniformly smooth Banach space. If $(x_i) \subset \mathfrak{X}$ is such that $\Sigma \|x_i\|^p < \infty$ and

$$\sum_{i=1}^{n_k} x_i \to x \in \mathfrak{X} , \qquad k \to \infty ,$$

for some $(n_k) \subset N$, then there exists a rearrangement σ of positive integers such that $\Sigma x_{\sigma(i)} \to x$.

Proof. The proof is based on the following inequality, valid in p - uniformly smooth Banach spaces, and which is an almost immediate corollary to the Theorem 2.1. :

there exists a constant $K > 0$ such that for any $n \in N$ and any $x_1,\dots,x_n \in \mathfrak{X}$ such that $x_1 + \dots + x_n = 0$ one can find a permutation $\sigma(1,\dots,n)$ such that for each k , $1 \leq k \leq n$

$$(*) \qquad \Big\| \sum_{i=1}^{k} x_{\sigma(i)} \Big\|^p \leq K \sum_{i=1}^{k} \|x_{\sigma(i)}\|^p .$$

Indeed, to prove it, it is enough to make an observation that given such $x_1,\dots,x_n$, and taking as a probability space $\Omega = \{\sigma : \sigma$ - permutation of $(1,\dots,n)\}$, equipped with uniform distribution, $x_{\sigma(1)},\dots,x_{\sigma(n)}$ become exchangeable random variables so that

$$Y_k(\sigma) = \frac{X_{\sigma(1)} + \dots + X_{\sigma(n-k)}}{n-k} , \qquad k = 0,1,\dots,n-1 ,$$

becomes a martingale with values in $\mathfrak{X}$ (relative to natural σ-algebras). Now, we apply the Theorem 2.1 to the norm super-martingale

$Y_k(n-k) = x_{\sigma(1)} + \dots + x_{\sigma(n-k)}$, and (after making an easy observation that if $\xi_1,\dots,\xi_n$ are adapted real random variables with $E(\xi_{i+1}/\Sigma_i) \geq 0$, $i = 0,\dots,n-1$ then one can find $w \in \Omega$ such that for every i $\xi_i(w) \geq 0$) get the existence of σ such that (*) is satisfied.

An obvious computation shows that for any $x_1,\dots,x_n \in \mathfrak{X}$ with $x_1 + \dots + x_n = s$ one can find a permutation σ of $(1,\dots,n)$ such that for each k , $1 \leq k \leq n$,

$$(**) \qquad \left\| \sum_{i=1}^{k} x_{\sigma(i)} \right\|^p \leq K \left(\|s\|^p + \sum_{i=1}^{k} \|x_{\sigma(i)}\|^p \right) ,$$

and, from the above inequality the theorem follows after rearranging the series Σx_i in blocks $(n_i, n_{i+1}-1)$ according to the permutations guaranteeing (**) . Q.E.D.

At this place we would like to make a loose digression and pose a question: how necessary is p - uniform smoothness (or smoothability) for validity of Theorem 2.5. ?

As the first idea one would want to prove that if in a Banach space $\mathfrak{X}$ the inequality (*) holds then $\mathfrak{X}$ can be equipped with an equivalent p - uniformly smooth norm for which the obvious candidate would be

$$\square\, x\, \square = \sup_{\Sigma x_i = x} \left(\frac{1}{c} \inf_{\sigma} \sup_{k} \left\| \sum_{i=1}^{k} x_{\sigma(i)} \right\|^p - \sum_{i=1}^{n} \|x_i\|^p \right)^{1/p} .$$

This is homogeneous, non-convex (but may be convexified), equivalent to $\|\cdot\|$ (also after convexification) so the only open question to be resolved is whether it gives rise to the p - uniformly smooth norm. ?

As another, and much deeper application, we indicate the theorem of Pisier [1] which says that every uniformly smooth (uniformly convex) Banach space may be renormed so that it becomes p - uniformly smooth (q - uniformly smooth) for some $p > 1$ $(q < \infty)$.

3. Super-properties of Banach spaces, finite tree property.

DEFINITION 3.1. (cf. Day [2]). Let $\underline{P}$ be a property of Banach spaces. We say that the Banach space $\mathfrak{X}$ has the property <u>super</u> $\underline{P}$ if any Banach space $\mathfrak{Y}$ which is finitely representable in $\mathfrak{X}$ has the property $\underline{P}$.

We recall that $\mathfrak{Y}$ is finitely representable in $\mathfrak{X}$ if for any $\varepsilon > 0$ and any finite-dimensional $\mathfrak{Y}^{fin} \subset \mathfrak{Y}$ one can find a finite dimensional subspace $\mathfrak{X}^{fin} \subset \mathfrak{X}$ such that

$$d(\mathfrak{X}^{fin}, \mathfrak{Y}^{fin}) \overset{df}{=}$$

$$\inf \left\{ \|T\| \, \|T^{-1}\| : T : \mathfrak{X}^{fin} \to \mathfrak{Y}^{fin} \text{ isomorphism} \right\} \leq 1 + \varepsilon .$$

Evidently,

$$\text{super } \underline{P} \Longrightarrow \underline{P} ,$$
$$\text{super (super } \underline{P}) \Longleftrightarrow \text{super } \underline{P}$$
$$(\underline{P} \Longrightarrow \underline{Q}) \Longrightarrow (\text{super } \underline{P} \Longrightarrow \text{super } \underline{Q}) .$$

Roughly speaking $\mathfrak{X} \in$ super $\underline{P}$ if any Banach space $\mathfrak{Y}$ the finite dimensional subspaces there of are similar to those of $\mathfrak{X}$ has $\underline{P}$.

Below, we collect some purely geometric characterizations of <u>super -reflexivity</u> needed for further reference. The notion was introduced by James [1]. At the beginning we recall a few notions that will appea in the following theorem.

DEFINITION 3.2. (James [1]). The Banach space $\mathfrak{X}$ has the <u>finite tree property</u> if there exists $\varepsilon > 0$ such that for any $n \in N$ one can find a binary tree $\left\{ x(\varepsilon_1,\dots,\varepsilon_k) : 1 \leq k \leq n, \ \varepsilon_i = \pm 1 \right\}$ contained in the unit ball of $\mathfrak{X}$ and such that

$$x(\varepsilon_1,\dots,\varepsilon_k) = \frac{1}{2} \left(x(\varepsilon_1,\dots,\varepsilon_k, 1) + x(\varepsilon_1,\dots,\varepsilon_k, -1) \right)$$

and

$$\| x(\varepsilon_1,\dots,\varepsilon_k) - x(\varepsilon_1,\dots,\varepsilon_k,\varepsilon_{k+1}) \| \geq \frac{\varepsilon}{2} ,$$

$k = 1,\dots,n-1$. (Essentially it is nothing else than finite Paley-Walsl martingale with uniformly big increments).

DEFINITION 3.3. A sequence $(x_n) \subset \mathfrak{X}$ is said to be <u>basic</u> with constant δ if for any scalar sequence $(\alpha_n) \subset R$ and all integers n,m

$$\delta \| \sum_{i=1}^{n} \alpha_i x_i \| \leq \| \sum_{i=1}^{n+m} \alpha_i x_i \| \quad .$$

THEOREM 3.1. *The following properties of a Banach space are equivalent:*

(a) $\mathfrak{X}$ *is super-reflexive* ,

(b) *there exists* $n \in N$ *and* $\varepsilon > 0$ *such that for all* $x_1, \ldots, x_n \in \mathfrak{X}$

$$\inf_{1 \leq k \leq n} \| \sum_{i=1}^{k} x_i - \sum_{i=k+1}^{n} x_i \| \leq n(1-\varepsilon) \sup_{1 \leq i \leq n} \|x_i\| \quad ,$$

(c) $\mathfrak{X}$ *does not possess the finite tree property,*

(d) *For each* $\delta > 0$ *there exists* $p > 1$ *and a constant* $C > 0$ *such that for any finite basic sequence* $(x_n) \subset \mathfrak{X}$ *with constant* δ

$$\|\Sigma x_n\| \geq C (\Sigma \|x_n\|^p)^{1/p} \quad ,$$

(e) *For each* $\delta > 0$ *there exists* $q < \infty$ *and a constant* $C > 0$ *such that for any finite basic sequence* $(x_n) \subset \mathfrak{X}$ *with constant* δ

$$\|\Sigma x_n\| \geq \frac{1}{C} (\Sigma \|x_n\|^q)^{1/q} \quad ,$$

(f) $L^2(\Omega, \mu; \mathfrak{X})$, $\mu(\Omega) > 0$, *is super-reflexive.*

The equivalence (a) <=> (b) is due to Schäffer and Sundaresan [1], (a) <=> (c) <=> (d) <=> (e) to James [1], [2] and (a) <=> (f) to Pisier [1].

Now, we are going to prove other characterizations of super-reflexivity in the language of $\mathfrak{X}$ - valued martingales. Recall, following Pisier [1], that any martingale on the probability space $\{-1,1\}^N$ equipped with Bernoulli probability and adapted to the σ-algebras generated by n-first variables is said to be Paley-Walsh martingale.

THEOREM 3.2. (Pisier [1], Kwapien-oral communication). *For the Banach space* $\mathfrak{X}$ *the following conditions are equivalent:*

(a) $\mathfrak{X}$ *is super-reflexive* ;

(b) $\mathfrak{X} \in$ *super* (MCT_p) ;

(c) $\mathfrak{X} \in$ *super* (RNP) ;

(d) <u>there exist a constant</u> $C > 0$ <u>and</u> $p > 1$ <u>such that for each</u> $\mathfrak{X}$ - <u>valued martingale</u> $(M_n) \subset L^2$

$$(*) \qquad \sup_n \|M_n\|_2 \leq C\,[\|M_0\|_2^p + \sum_{n=0}^{\infty} \|M_{n+1} - M_n\|_2^p]^{1/p}$$

<u>where</u>

$$\|M\|_2 \overset{\text{df}}{=} (E\|M\|^2)^{1/2} \quad ;$$

(e) <u>there exist a constant</u> $C > 0$ <u>and</u> $p > 1$ <u>such that for each</u> $\mathfrak{X}$ - <u>valued Paley-Walsh martingale</u> (M_n) $(*)$ <u>holds true</u> ;

(f) <u>there exist a constant</u> $C > 0$ <u>and</u> $q < \infty$ <u>such that for each</u> $\mathfrak{X}$ - <u>valued martingale</u> $(M_n) \subset L^2$

$$(**) \qquad (\|M_0\|_2^q + \sum_{n=0}^{\infty} \|M_{n+1} - M_n\|_2^q)^{1/q} \leq C \sup_n \|M_n\|_2 \quad ;$$

(g) <u>there exist a constant</u> $C>0$ <u>and</u> $q < \infty$ <u>such that for each</u> $\mathfrak{X}$ - <u>valued Paley-Walsh martingale</u> (M_n) $(**)$ <u>holds true</u>.

<u>Proof</u>. We prove the following implications: (a)=> (b) => (a) , (b) <==> (c) , (a) => (f) => (g) => (a) , (a) => (d) => (e) => (g) .

(a) => (b) follows from the fact that super-reflexive spaces are reflexive and that reflexive spaces have (MCT_p) (cf. Section 1).

(b) => (a). Assume that $\mathfrak{X}$ is not super-reflexive i.e. there exists a Banach space $\mathfrak{Y}$ which is finitely representable in $\mathfrak{X}$ and which has an infinite ε - tree in its unit ball for some $\varepsilon > 0$. But the same tree may be looked at as a bounded Paley-Walsh martingale with values in $\mathfrak{Y}$, which does not converge a.s. This evidently contradicts (b).

(b) <==> (c) follows from Theorem 1.1.

(a) => (d) and (a) => (f) follow directly from the Theorem 3.1 because the increments of square integrable martingale form a basic sequence in $L^2(\Omega,\Sigma,P;\ \mathfrak{X})$ with constant 1. Indeed, for any $(\alpha_i) \subset R$

$$(E\,\|\sum_{i=1}^{n+m} \alpha_i\,(M_i-M_{i-1})\|^2)^{1/2} = [E\,E\,(\|\sum_{i=1}^{n+m} \alpha_i(M_i-M_{i-1})\|^2/\Sigma_n)]^{1/2} \geq$$

$$\geq [E\|E(\sum_{i=1}^{n+m} \alpha_i(M_i-M_{i-1})/\Sigma_n)\|^2]^{1/2} = [E\|\sum_{i=1}^{n} \alpha_i(M_i-M_{i-1})\|^2]^{1/2} .$$

(d) => (e) and (f) => (g) are obvious.

(g) => (a) Assume that $\mathfrak{X}$ is not super-reflexive. Then it has the finite tree property by Theorem 3.1., that is there exists an $\varepsilon > 0$ such that for all $n \in N$ there exists a Paley-Walsh martingale (M_n) of length n, with values in the unit ball of $\mathfrak{X}$ such that

$$\|M_{n+1} - M_n\| \geq \varepsilon/2 .$$

Thus, were the inequality (**) satisfied, we would have that

$$1 \geq (E\|M_n\|^2)^{1/2} \geq \frac{1}{C} [\sum_{k=1}^{n} \|M_n-M_{n-1}\|_2^q]^{1/q} \geq (\frac{\varepsilon}{2C})n^{1/q} ,$$

a contradiction.

(e) => (g). We prove this implication for $\mathfrak{X}^*$ but that suffices because, by the implication (g) => (a) we would have that $\mathfrak{X}^*$ is super-reflexive and so is $\mathfrak{X}$ (super-reflexivity is the self-dual property by James [1]).

So let (M_n^*) be a Paley-Walsh martingale with values in $\mathfrak{X}^*$. Then

$$(\|M_o\|_2^q + \sum_{k=1}^{n} \|M_k^* - M_{k-1}^*\|_2^q)^{1/q}$$

$$= \sup \left\{E\, M_o^* X_o + \sum_{k=1}^{n} E(M_k^* - M_{k-1}^*)X_k : \|X_o\|_2^p + \sum_{k=1}^{n} \|X_k\|_2^p \leq 1,\ X_i \in L^p(\mathfrak{X})\right\}$$

$$= \sup E\, [M_n^*\, E(X_o/\Sigma_o)] + \sum_{k=1}^{n} E\, [M_n^*\, (E(X_k/\Sigma_k) - E(X_k/\Sigma_{k-1}))]$$

$$\leq \|M_n^*\|_2 \sup \|E(X_o/\Sigma_o) + \sum_{k=1}^{n} [E(X_k/\Sigma_k) - E(X_k/\Sigma_{k-1})]\|_2$$

$$\leq C\|M_n^*\|_2 \sup(\|E(X_o/\Sigma_o)\|_2^p + \sum_{k=1}^{n} \|E(X_k/\Sigma_k) - E(X_k/\Sigma_{k-1})\|_2^p)^{1/p} \leq 2C\|M_n^*\|_2$$

what ends the proof.

Remark 3.1. Enflo [1] has shown that the super-reflexivity is equivalent to uniform convexifiability and this, in turn, is equivalent to uniform smoothability. Actually, by Asplund's result, in this case the space may be equipped with an equivalent norm which is at the same time uniformly smooth and uniformly convex. Pisier [1], using in the essential manner, the martingale technique developed above that uniform convexifiability is equivalent to q - uniform convexifiability for some $q < \infty$. For more results on uniform convexifiability see Beauzamy [1], [2].

4. Dentability, lattice bounded operators, and submartingales in vector lattices.

In this section we present, essentially, the results contained in the paper by Szulga and Woyczyński [1] concerning the convergence theorems for submartingales taking values in vector lattices. For real -valued submartingales there exists the following fundamental

PROPOSITION 4.1. (cf. e.g. Neveu [1], p. 63). *If $(X_n, n \in N)$ is a real-valued submartingale, and $\sup_n EX_n^+ < \infty$ then there exists an $X_\infty \in L^1$ such that $X_n \to X_\infty$ a.s.*

What we are interested in here, is how the Theorem 4.1 (which can also be dually formulated for supermartingales) carries over to the case of submartingales with values in Banach lattices (to be defined below). In what we did we were encouraged by the fact that recently Schwartz [1] extensively developed the theory of supermartingales that have measures as their values and applied it efficiently to the desintegration of measures. His model fits into our general framework.

DEFINITION 4.1. The vector lattice $(\mathfrak{X},\leq)$ is said to be a *Banach lattice* if it is equipped with monotone ($|x| \leq |y|$ implies $\|x\| \leq \|y\|$) and complete norm.

As usual, if $x \in \mathfrak{X}$ then $x^+ \overset{df}{=} \sup(x,0)$, $x^- \overset{df}{=} \sup(-x,0)$, $|x| \overset{df}{=} x^+ + x^-$. $\mathfrak{X}^*$, the norm dual of $\mathfrak{X}$, is also a Banach lattice lattice under the natural ordering, and by $\mathfrak{X}_+$ and $\mathfrak{X}_+^*$ we denote non-negative cones in $\mathfrak{X}$ and $\mathfrak{X}^*$, respectively.

DEFINITION 4.2. The set $A \subset \mathfrak{X}$ is said to be <u>order bounded</u> if there exists $x_0 \in \mathfrak{X}$ such that for all $y \in A$, $|y| \leq x_0$. The linear operator T from a Banach space $\mathfrak{Y}$ into a Banach lattice $\mathfrak{X}$ is called <u>lattice bounded</u> if it maps the unit ball of $\mathfrak{Y}$ into an order bounded subset of $\mathfrak{X}$.

Recently, the papers by Garling [1], Nielsen [1] and others (Kwapień, Vershik, Sudakov) raised the interest in such operators in connection with absolutely summing and radonifying operators, however, somehow **neglecting** the old results of Kantorovich, Vulikh and Pinsker [1], (VIII.4 - VIII.6) quoted below.

PROPOSITION 4.2. (a) <u>Let C be the Banach space of real continuous functions on the unit interval and let $\mathfrak{X}$ be a separable Banach lattice with order continuous norm. Then $T : C \to \mathfrak{X}$ is lattice bounded if, and only if there exists $g : [0,1] \to \mathfrak{X}$ of bounded o-variation</u>

$$\text{ess var } g(t) \overset{df}{=} \sup \Sigma \, |g(t_{i+1}) - g(t_i)| \in \mathfrak{X}$$

<u>such that</u>

$$T f = \int_0^1 f(t) \, dg(t) , \qquad f \in C ,$$

<u>where the integral is understood as an order limit of Stieltjes sums:</u>

(b) <u>The operator $T : l^q \to l^p$ $(q > 1, p \geq 1)$ is lattice bounded if, and only if it is of the form</u>

$$T y = \Big(\sum_{k=1}^{\infty} a_{ik} y_k \Big)_{i \in N} , \qquad y = (y_k) \in l^q ,$$

<u>where</u>

$$\sum_{i=1}^{\infty} \Big[\sum_{k=1}^{\infty} |a_{ik}|^{q/(q-1)} \Big]^{(q-1)p/q} < \infty .$$

(c) <u>The operator $T : \mathfrak{Y} \to L^p[0,1]$, $(p \geq 1)$ where $\mathfrak{Y}$ is a separable Banach space is lattice bounded if, and only if it is of the form</u>

$$(Ty)(t) = f^*(t)y \quad ,$$

$y \in \mathfrak{Y}$, $t \in [0,1]$, <u>and</u> $f^* : [0,1] \to \mathfrak{Y}$ <u>is</u> $*$-<u>weakly measurable and such that</u> $\|f^*\| \in L^p[0,1]$.

Now, we shall formulate the result on dentable Banach lattices that will be used later on.

PROPOSITION 4.3. <u>If</u> $\mathfrak{X}$ <u>is the dentable Banach lattice and the sequence</u> $x_0 \leq x_1 \leq x_2 \ \ldots$ <u>is norm bounded then it is convergent</u>.

<u>Proof</u>. Because $\mathfrak{X}$ is dentable it does not contain isomorphic copies of c_0 (Example 1.2 and Theorem 1.1) and in every such Banach lattice monotone norm-bounded sequences are convergent by Tzafriri [1], Th. 14 .

DEFINITION 4.3. Let (Ω,Σ,P) be a probability space and let $\Sigma_1 \subset \Sigma_2 \subset \Sigma_3 \subset \ldots \subset \Sigma$ be a sequence of sub-σ-algebras. The sequence $(X_n,\Sigma_n,\ n \in N)$, $X_n \in L^1(\Omega,\Sigma_n,P;\ \mathfrak{X})$, where $\mathfrak{X}$ is a vector lattice is said to be a <u>sub-martingale</u> cf $E(X_{n+1}/\Sigma_n) \geq X_n$, a.s.

Now, we turn to the investigation of analogues of Proposition 4.1 for $\mathfrak{X}$ - valued sub-martingales.

Notice that the Doob's condition $\sup_n EX_n^+ < \infty$ for real-valued random variables has two analogues for Banach-lattice-valued random vectors, namely: order boundedness of $(E(X_n^+),\ n \in N)$ and $\sup_n E\|X_n^+\| < \infty$ .Both boil down to the Doob's condition in the real case however, as we shall see below, in general, neither is sufficient to assure the a.s. convergence of a submartingale $(X_n, n \in N)$.

It is not difficult to check that for both, real and vector sub-martingales, the set $(E(X_n^+),\ n \in N)$ is order bounded if, and only if $(E|X_n|,\ n \in N)$ is such. However, even for vector-valued <u>martingales</u> it might happen that $\sup_n E\|X_n^+\| < \infty$ and still $\sup_n E\|X_n^-\| = \infty$ so that it will be not surprising that the condition $\sup_n E\|X_n^+\| < \infty$ does not, in general, imply the a.s. convergence of a submartingale (X_n) even in dentable Banach lattices. On the other hand the condition $\sup_n E\|X_n^+\| < \infty$ is stronger than order boundedness of $(E(X_n^+),\ n \in N)$ for any sequence (X_n) of random vectors with values in the Banach lattice $\mathfrak{X}$ because for each $x^* \in \mathfrak{X}_+^*$, $\sup_n x^*E(X_n^+) \leq \sup_n E\|X_n^+\|\|x^*\| <$ $< \infty$, and because the set $A \subset \mathfrak{X}$ is order bounded if, and only if for each $x^* \in \mathfrak{X}^*_+$, x^*A is bounded on the real line. The last statement follows from the fact that in a Banach lattice $\mathfrak{X}$, $x \geq 0$ if, and only if for each $x^* \in \mathfrak{X}^*_+$, $x^*x \geq 0$.

It is not hard to see that if $\mathfrak{X}$ is a dentable Banach lattice then in order to produce examples of

(e) a martingale (M_n) with values in $\mathfrak{X}$ such that $\sup_n E\|M_n^+\| < \infty$ and at the same time $\sup_n E\|M_n^-\| = \infty$ and $\sup_n E\|M_n\| = \infty$, and of

(ee) a submartingale (X_n) with values in $\mathfrak{X}$ such that $\sup_n E\|X_n^+\| < \infty$ and X_n diverges a.s, it is sufficient to find

(eee) an a.s. divergent sequence (Y_n) of non-negative independent random vectors in $L^1(\Omega,\Sigma,P;\mathfrak{X})$ such that both $\sup_n E\|Y_n\| < \infty$ and $\sup_n \|\sum_{i=0}^{n} E\, Y_i\| < \infty$.

Indeed, given such a sequence (Y_n) it is enough to take $\Sigma_n = \sigma(Y_0,\ldots,Y_n)$, $Z_0 = 0$,

$$Z_n = \sum_{i=0}^{n-1} Y_i \;, \qquad M_n = -Z_{n+1} + E\, Z_{n+1} \;,$$

$$X_n = M_n + Z_n = E\, Z_{n+1} - Y_n \;, \qquad n \geq 1 \;.$$

Then the sequence (EZ_n) converges because of Proposition 4.3, and on the other hand the sequence (Z_n) diverges a.s. because the sequence (Y_n) itself diverges so that $\sup_n E\|Z_n\| = \infty$.

Now, (M_n) defined in such a way is a zero mean martingale such that $\sup_n E\|M_n^+\| \leq \sup_n \|EZ_{n+1}\| < \infty$ (because $M_n^+ \leq EZ_{n+1}$) but, at the same time

$$E\|M_n\| = E\,\|Z_{n+1} - E\, Z_{n+1}\| \geq E\|\,Z_{n+1}\| - \|E\, Z_{n+1}\|$$

is unbounded. X_n is evidently a submartingale that is divergent a.s. and for which

$$\sup_n E\|X_n^+\| \leq \sup_n \|E(Z_{n+1})\| = \sup_n \|E\, Z_{n+1}\| < \infty$$

Given below are examples of sequences (Y_n) of random vectors with values in certain classical Banach lattices that enjoy the property (eee).

Example 4.1. Let $\mathfrak{X} = l^p$, $p > 1$ (reason why $p = 1$ is excluded appears in Corollary at the end of this section), $\Omega_i = [0,1)$, Σ_i be all Borel subsets of $[0,1)$, and λ_i be the Lebesgue measure, $i \in N$. Put

$$\Omega = \Pi_{i\in N}\,\Omega_i \quad , \quad \Sigma = \Pi_{i\in N}\,\Sigma_i \quad , \quad P = \Pi_{i\in N}\,\lambda_i \quad ,$$

$$Y_0(w_0,w_1,\dots) = 0 \quad ,$$

and

$$Y_{2^{n-1}+k}(w_0,w_1,\dots) = I_{[k/2^{n-1},(k+1)/2^{n-1})}(w_{2^{n-1}+k})\; e_{2^{n-1}+k} \quad ,$$

where $n = 1,2,\dots$, $k = 0,1,\dots,2^{n-1}-1$, I_A is the indicator function of the set A and $(e_n,\ n \in N)$ is the standard basis in l^p.

By definition $(Y_n,\ n \in N)$ are independent, non-negative and in $L^1(\Omega,\Sigma,P;\ \mathfrak{X})$, $\sup_{n\in N} E\|Y_n\| \leq 1$ and

$$\sup_{n\in N} \|\sum_{i=1}^{n} E\, Y_i\| < \infty$$

because

$$EY_0 = 0 \quad , \quad EY_{2^{n-1}+k} = 2^{1-n}\, e_{2^{n-1}+k} \quad ,$$

$n = 1,2,\dots$, $k = 0,1,\dots,2^{n-1}-1$, so that

$$\sup_{n\in N} \|\sum_{i=0}^{n} EY_i\| = \|\sum_{i=0}^{\infty} EY_i\| = (\sum_{i=0}^{\infty} 2^{i(1-p)})^{1/p} \quad ,$$

but, at the same time, (Y_n) is divergent for each $w \in \Omega$ because for each $w \in \Omega$ there exist sequences (n_i) , $(n_i') \subset N$ such that $\|Y_{n_i}(w)\| = 1$ and $\|Y_{n_i'}(w)\| = 0$.

Remark 4.1. Having written down the above example we can't resist quoting the following remark made by J.Szulga (oral communication).

If $\mathfrak{X}$ is a separable Banach space and $(x_n) \subset \mathfrak{X}$ then a formulation of Banach-Steinhaus theorem claims that if for each $x^* \in \mathfrak{X}^*$, $\sup_n |x^* x_n| < \infty$ then $\sup_n \|x_n\| < \infty$. C.Ryll-Nardzewski asked whether the random version of this theorem is true i.e. if the a.s. finiteness of $\sup_n |x^* X_n|$ for any $x^* \in \mathfrak{X}$ quarantees that $\sup_n \|X_n\| < \infty$ a.s, (X_n) being the $\mathfrak{X}$ - valued random vectors. The Example 4.1 provides the negative answer to this question. Indeed, take a sequence $(Z_n, n \in N)$ from the reasoning preceding Example 4.1. $\sup_n \|Z_n\| = \infty$ a.s. and at the same time. $EZ_n \leq z \in \mathfrak{X}$ so that for each $x^* \geq 0$, $Ex^* Z_n \leq x^* z$. Hence, by Lebesgue monotone convergence theorem $E \sup_n x^* Z_n \leq x^* z$ and $\sup_n x^* Z_n < \infty$ a.s. for all $x^* \geq 0$. Because in every Banach lattice $x^* = x^*_+ - x^*_-$ where both $x^*, x^*_- \geq 0$ we get the necessary counter-examples for all $\mathfrak{X} = l_p$, $p > 1$.

Now, we pass to the positive results concerning the convergence of submartingales. We start with an observation that the Doob's decomposition of a real submartingale survives in a Banach lattice. Namely, if $(X_n, \Sigma_n, n \in N)$ is a submartingale with values in a Banach lattice $\mathfrak{X}$ then

$$X_n = M_n + Z_n ,$$

where (M_n, Σ_n) is a martingale, and the sequence $(Z_n, n \in N)$ is predictable (i.e $Z_n \in L^1(\Omega, \Sigma_{n-1}, P; \mathfrak{X})$ and such that $A_0 = 0$, $0 \leq A_n \uparrow$ a.s. (the decomposition is unique). As in the real case, to prove the above statement it is sufficient to put $A_0 = 0$, $M_0 = X_0$,

$$M_n = X_0 + \sum_{i=1}^{n} [X_i - E(X_i/\Sigma_{i-1})] ,$$

$$Z_n = \sum_{i=1}^{n} E(X_i - X_{i-1}/\Sigma_{i-1}) , \qquad n = 1,2,\dots .$$

Example 3.1 shows that, in general, the condition $\sup_n E\|X_n^+\| < \infty$, for a submartingale $X_n = M_n + Z_n$, does not imply its a.s. convergence. However, we do have.

THEOREM 4.1. (Szulga and Woyczyński [1]). <u>For the separable Banach lattice</u> $\mathfrak{X}$ <u>the following conditions are equivalent:</u>

(α) $\mathfrak{X}$ <u>is dentable</u> ;

(β) <u>for each $\mathfrak{X}$ - valued submartingale</u> $(X_n = M_n + Z_n, n \in N)$ <u>satisfying the conditions</u> $\sup_n E\|X_n^+\| < \infty$ <u>and</u> $\sup_n E\|M_n^-\| < \infty$ <u>there exists an</u> $X_\infty \in L^1(\Omega,\Sigma,P; \mathfrak{X})$ <u>such that</u> $X_n \to X_\infty$ a.s.;

(γ) <u>for each $\mathfrak{X}$ - valued submartingale</u> $(X_n = M_n + Z_n, n \in N)$ <u>satisfying the conditions</u> $\sup_n E\|X_n^+\|^p < \infty$ <u>and</u> $\sup_n E\|M_n^-\|^p < \infty$ <u>for some</u> $1 < p < \infty$ <u>there exists an</u> $X_\infty \in L^p(\Omega,\Sigma,P; \mathfrak{X})$ <u>such that</u> $X_n \to X_\infty$ <u>a.s. and in</u> L^p .

<u>Proof.</u> (α) => (β) [(α) => (γ)]. $X_n = M_n + Z_n \geq M_n$ a.s. so that $X_n^+ \geq M_n^+$ a.s. and the monotonicity of the norm yields that

$$\sup_{n\in N} E\|M_n\|^p \leq 2^p(\sup_{n\in N} E\|X_n^+\|^p + \sup_{n\in N} E\|M_n^-\|^p) .$$

Hence, by Theorem 1.1 there exists $M_\infty \in L^1$ $[M_\infty \in L^p]$ such that $M_n \to M_\infty$ a.s. $[M_n \to M_\infty$ a.s. and in $L^p]$. Because $Z_n = X_n - M_n$, we have that $Z_n \leq X_n^+ + M_n^-$ so that $\sup_{n\in N} E\|Z_n\|^p < \infty$, $1 \leq p < \infty$. Utilizing again the monotonicity of the norm and the Lebesgue monotone convergence theorem we get that $E \sup_{n\in N}\|Z_n\|^p < \infty$ so that $\sup_{n\in N}\|Z_n\| < \infty$ a.s. However, because of Proposition 4.3 there exists a random vector Z_∞ such that $Z_n \to Z_\infty$ a.s. The Fatou lemma yields that

$$E\|Z_\infty\|^p \leq \liminf_{n\in N} E\|Z_n\|^p \leq \sup_{n\in N} E\|Z_n\|^p < \infty$$

so that $Z_\infty \in L^p$, $1 \leq p < \infty$, and letting

$$X_\infty = M_\infty + Z_\infty$$

ends the proof.

(β) => (α) [(γ) => (α)]. If $X_n = M_n$ then the conditions in (β) [(γ)] boil down to $\sup_{n\in N} E\|M_n\| < \infty$ $[\sup_{n\in N} E\|M_n\|^p < \infty]$ and the Theorem 1.1 gives the dentability of $\mathfrak{X}$.

In the next theorem we make weaker assumptions about the submartingale $(X_n, n \in N)$, namely that $(E(X_n^+), n \in N)$ is order bounded, but the convergence takes place only for a transformed submartingale.

THEOREM 4.2. (Szulga and Woyczyński [1]). <u>Let</u> $\mathfrak{X}$ <u>be a separable Banach lattice,</u> $\mathfrak{Y}$ <u>a dentable separable Banach lattice, and</u> $T : \mathfrak{X} \to \mathfrak{Y}$

<u>a linear bounded positive operator such that its transpose</u> $T^* : \mathfrak{Y}^* \to \mathfrak{X}^*$ <u>is lattice bounded. If</u> $(X_n, n \in N)$ <u>is a submartingale with values in</u> $\mathfrak{X}$ <u>such that</u> $(E(X_n^+), n \in N)$ <u>is order bounded then there exists a</u> $Y_\infty \in L^1(\mathfrak{Y})$ <u>such that the submartingale</u> $TX_n \to Y_\infty$ <u>a.s.</u> $n \to \infty$.

Proof. Let $X_n = M_n + Z_n$ as before, and let $E(X_n^+) \leq x_0 \in \mathfrak{X}_+$ for all $n \in N$. We show that under the above assumption

$$\sup_{n\in N} E\|(TX_n)^+\| < \infty \quad ,$$

and

$$\sup_{n\in N} E\|(TM_n)^-\| < \infty \quad ,$$

what, in view of the Theorem 4.1 would give the desired result because the Doob's decomposition for the submartingale TX_n is $TM_n + TZ_n$. Indeed,

$$\sup_{n\in N} E\|(TX_n)^+\| \leq \sup_{n\in N} E\|T(X_n^+)\|$$

$$= \sup_{n\in N} E \sup\left\{y^* T(X_n) : \|y^*\| \leq 1,\ 0 \leq y^* \in \mathfrak{Y}^*\right\}$$

$$= \sup_{n\in N} E \sup\left\{(T^* y^*)X_n^+ : \|y^*\| \leq 1,\ 0 \leq y^* \in \mathfrak{Y}^*\right\} .$$

However, the transpose T^* of a positive T is also positive, and thus the fact that T^* is lattice bounded implies the existence of an $x_0^* \in \mathfrak{X}_+^*$ such that for each y^* with $\|y^*\| \leq 1$, $|T^* y^*| \leq x_0^*$. Thus we get that

$$\sup_{n\in N} E\|(TX_n)^+\| \leq E(x_0^* X_n^+) = x_0^* E(X_n^+)$$

$$\leq x_0^* x < \infty \quad .$$

Proceeding as above and utilizing the inequality

$$E(TM_n)^- = E(TM_n)^+ - E(TM_n) = E(TM_n)^+ - E(TM_0) \leq E(TX_n)^+ - E(TM_0)$$

we get

$$\sup_{n\in N} E\|(TM_n)^-\| \leq x_0^* x_0 + |x_0^* E(TM_0)| < \infty$$

because $M_0 \in L^1$. This ends the proof.

Because l^1 is the dentable Banach lattice (cf. Proposition 1.3 and preceding comments), and because the operator $[Id(l^1,l^1)]^*$ is lattice bounded in l^∞ we obtain the following.

COROLLARY 4.1. _If_ $(X_n, n \in N)$ _is a submartingale with values in_ l^1 _such that_ $(E(X_n^+), n \in N)$ _is order bounded then there exists an_ $X_\infty \in L^1(\Omega,\Sigma,P; l^1)$ _such that_ $X_n \to X_\infty$ _a.s._

References

E.ASPLUND
[1] Averaged norms, Israel J. Math. 5(1967), 227-233.

E.ASPLUND and I.NAMIOKA
[1] A geometric proof of Ryll-Nardzewski's fixed point theorem, Bull. Amer. Math. Soc. 73(1967), 443-445.

P.ASSOUAD
[1] Martingales et rearrangement dans les espaces uniformement lisses, Preprint, Orsay, Julliet 1974.

B.BEAUZAMY
[1] Espaces de Banach uniformément convexifiables, Seminaire Maurey-Schwartz 1973-74, XIII.1-18, XIV.1-17.
[2] Operateurs uniformement convexifiables, Preprint, Centre de Math., Ecole Polytechnique, Paris, Octobre 1974.

S.D.CHATTERJI
[1] Martingales of Banach-valued random variables, Bull. Amer. Math. Soc. 66(1960), 395-398.
[2] A note on the convergence of Banach-space valued martingales, Math. Ann. 153(1964), 142-149.
[3] Martingale convergence and the Radon-Nikodym theorem in Banach spaces, Math. Scand. 22(1968), 21-41.

J.A.CLARKSON
[1] Uniformly convex spaces, Trans. Amer. Math. Soc. 40(1936), 396-414.

W.J.DAVIS
[1] The Radon-Nikodym property, Seminaire Maurey-Schwartz 1973-74, 0.1-0.12.

W.J.DAVIS and R.R.PHELPS
[1] The Radon-Nikodym property and dentable sets in Banach spa-

ces, Proc. Amer. Math. Soc. 45(1974), 119-122.

M.M.DAY

[1] Uniform convexity in factor and cojugate spaces, Ann. Math. 45(1944), 375-385.

[2] Normed linear spaces, Third Edition, Springer-Verlag, Berlin-Heidelberg-New York 1973.

J.DIESTEL

[1] About Radon-Nikodym Property and Radon-Nikodym Theorem, unpublished notes, Kent State U 1973.

J.DIESTEL and J.J.UHL,JR.

[1] The Radon-Nikodym Theorem for Banach space valued measures, Rocky Mountains J., to appear.

N.DUNFORD and J.T.SCHWARTZ

[1] Linear operators, Part I : General theory, Interscience, New York 1958.

P.ENFLO

[1] Banach spaces which can be given an equivalent uniformly convex norm, Israel J. Math. 13(1972), 281-288.

T.FIGIEL

[1] On the moduli of convexity and smoothness, Studia Math., to appear.

D.J.H.GARLING

[1] Lattice bounding, radonifying and summing mappings, 1974, to appear.

J.HOFFMANN - JØRGENSEN

[1] On the modulus of smoothness and the G_α-conditions in B-spaces, Aarhus Universitet, Matematisk Institut, Preprint Series, September 1974.

[2] Sums of independent Banach space valued random variables, Aarhus Universitet, Preprint series No 15, 1972/1973.

[3] The strong law of large numbers and the central limit theorem in Banach spaces, Aarhus Universitet, Matematisk Institut, Preprint Series, September 1974.

R.E.HUFF

[1] Dentability and the Radon-Nikodym Property, Duke Math. J. 41(1974), 111-114.

R.E.HUFF and P.D.MORRIS

[1] Geometric characterization of the Radon-Nikodym Property in Banach spaces, Studia Math. (to appear).

[2] Dual spaces with the Krein-Milman property, Proc. Amer. Math. Soc. (to appear).

R.C.JAMES

[1] Some self-dual properties of normed linear spaces, Ann. of Math. Studies 69(1972), 159-176.

[2] Super-reflexive spaces with bases, Pacific J. Math.41(1972), 409-420.

M.I.KADEC

[1] On conditionally convergent series in spaces L^p , Uspekhi Mat. Nauk 9(1)(1954), 107-109.

L.V.KANTOROVICH, B.Z.VULIKH and A.G.PINSKER
[1] Functional analysis in semiordered spaces, (in Russian), Moscow 1950.

J.KELLEY, I.NAMIOKA, et al.
[1] Linear topological spaces, Van Nostrand, New York 1963.

D.R.LEWIS
[1] A vector measure with no derivative, Proc. Amer. Math. Soc. 32(1972), 535-536.

J.LINDENSTRAUSS
[1] On the modulus of smoothness and divergent series in Banach spaces, Michigan Math. J. 10(1963), 241-252.

B.MAUREY and G.PISIER
[1] Series de variables aleatoires vectorielles independantes et proprietes geometriques des espaces de Banach, Mimeographed notes, Ecole Polytechnique, Paris, November 1974.

H.B.MAYNARD
[1] A gepmetrical characterization of Banach spaces with the Radon-Nikodym Property, Trans. Amer. Math. Soc. 185(1973), 493-500.

M.METIVIER
[1] Limites projectives de mesures; martingales; applications, Ann. Math. Pura Appl. 63(1963), 225-352.

V.D.MILMAN
[1] Geometric theory of Banach spaces, Part. II: Geometry of the unit sphere, Uspekhi Mat. Nauk 26(6)(1971), 73-149.

K.MUSIAŁ, C.RYLL-NARDZEWSKI and W.A.WOYCZYŃSKI
[1] Convergence presque sure des séries aléatoires vectorielles à multiplicateur bornés, C.R. Acad. Sci. Paris 279(1974), 225-228.

J.NEVEU
[1] Martingales à temps discret, Masson and Cie, Paris 1972.

N.J.NIELSEN
[1] On Banach ideals determined by Banach lattices and their applications, Dissertationes Math. 109(1973), 1-66.

G.PISIER
[1] Martingales à valeurs dans les espaces uniformement convexes; Handwritten mimeographed notes, Ecole Polytechnique, Paris, June 1974.

R.R.PHELPS
[1] Dentability and extreme points in Banach spaces, J.Functional Anal. 16(1974), 78-90.

M.A.RIEFFEL
[1] Dentable subsets of Banach spaces with application to a Radom-Nikodym Theorem, Functional Analysis, Thompson Book Co 1967, 71-77.
[2] The Radom-Nikodym Theorem for the Bochner integral, Trans. Amer. Math. Soc. 131(1968), 466-487.

F.S.SCALORA
[1] Abstract martingale convergence theorem, Pacific J. Math. 11(1961), 347-374.

J.J.SCHAFFER and K.SUNDARESAN

[1] Reflexivity and the girth of spheres, Math. Ann. 184(1970), 163-168.

L.SCHWARTZ

[1] Surmartingales régulières à valeurs mesures et désintégrations régulières d'une mesure, J. Anal. Math. (Jerusalem) 26(1973), 1-168.

J.SZULGA and W.A.WOYCZYŃSKI

[1] Convergence of submartingales in Banach lattices, 1974, to appear.

A.I.TULCEA and C.I.TULCEA

[1] Abstract ergodic theorems, Proc. Nat. Ac. Sci. 48(1962), 204-206.

L.TZAFRIRI

[1] Reflexivity in Banach lattices and their subspaces, J.Functional. Anal. 10(1972), 1-18.

J.J.UHL, JR

[1] A note on the Radon-Nikodym Property for Banach spaces, Rev. Roum, Mat. 17(1972), 113-115.

W.A.WOYCZYŃSKI

[1] Random series and laws of large numbers in some Banach spaces, Teorya Veroyatnostej i Prim. 18(1973), 371-377.

[2] Strong laws of large numbers in certain linear spaces, Ann. Inst. Fourier 24(1974), 205-223.

WINTER SCHOOL
ON PROBABILITY
Karpacz 1975
Springer's LNM

A NOTE ON SEMIPOLAR SETS FOR PROCESSES WITH INDEPENDENT INCREMENTS

By J. Zabczyk
Institute of Mathematics , Polish Academy of Sciences.

1. Let $X = (\Omega, \mathcal{M}, \mathcal{M}_t, X_t, \theta_t, P^x)$ be a standard Markov process on a state space E . With every such process it is possible to associate some classes of "exceptional" sets: the so called polar, semipolar and null sets which play an important role in probabilistc potential theory. For the definitions of exceptional sets as well as for the definitions of standard Markov processes, state space, potentials and other related notions we refer to the monograph [1]. Polar sets are semipolar and every semipolar set in null. The property: "every semipolar set is polar" is the hypothesis (H) of Hunt's paper ([4], p. 193) and it holds for a large class of Markov processes. For instance the hypothesis (H) is satisfied for symmetric processes with independent increments which have a reference measure (e.g. for Brownian motion, symmetric stable processes). In this note we prove that for processes with independent increments for which there exists a reference measure a similar property:

"every null Borel set is semipolar"

is never satisfied. Thus such processes are not "bizarres" (strange) in Dellacherie's sense [2]. In, this way we also answer a question posed in [5, p. 235].

2. Let R and T denote the real line and the unit circle respectively treated as groups. We recall that a Markov process X on

the state space $E = R^d$ or T^d is called to be a process with independent increments (shortly p.w.i.i.) if its transition function $P_t(\cdot,\cdot)$ is given by the formula:

$$P_t(x,A) = \mu_t(A-x) \ , \quad x \in E \ , \quad A \text{ any Borel subset of } E \ ,$$

where $(\mu_t)_{t\geq 0}$ is a weakly continuous semigroup (under convolution) of probabilistic measures on E such that μ_0 is unit mass at the origin. By ξ_d we shall denote the Haar measure on $E = R^d$ or $E = T^d$ and if the value of d is clear from the context we shall write simply ξ.

Fubini's theorem implies that if A is null for a process w.i.i. X then $\xi(A) = 0$. Thus null sets are rather small. On the other hand if $d \geq 2$ and there exists a reference measure for X then every countable subset of E is polar and therefore semipolar for X. Moreover if A is the union of any countable family of spheres with the centres at the origin then there exists an isotropic p.w.i.i. for which A is a semipolar (even a polar) set. Thus the family of semipolar sets is sometimes rather rich. It turns out however that the family of all Borel semipolar sets is essentially smaller than the family of all Borel sets ξ - negligible. This follows from the following theorem:

THEOREM 1. *If X is a precess w.i.i. then there exists a compact set $K \subset E$ such that $\xi(K) = 0$ and K is not semipolar for X.*

3. The proof of the theorem 1 will be based on the lemma below, the proof whereof is due to J.P. Kahane.

LEMMA. *If $u \in L^1(T^d)$ then there exists a singular (with respect to ξ_d) probability measure μ on T^d such that the convolution $u * \mu$ is, after a modification on a ξ_d - negligible set, a continuous function.*

Proof. Let $d = 1$. Define measure μ as a Riesz product (for the definition see [6,V])

$$\frac{1}{2\pi} \prod_{n=1}^{+\infty} (1 + \cos \lambda_n x)$$

where $x \in [-\pi,\pi)$ and λ_n are positive integers such that $\lambda_{n+1}/\lambda_n \geq q > 3$ and

$$\sup_{|k|\geq\lambda_n/2} |\hat{u}(k)| \leq 2^{-n}\, n^{-2} .$$

Here $\hat{u}$ denotes the Fourier's transform of the function u . Since $u \in L^1(T^d)$ thus $|\hat{u}(k)| \to 0$ as $|k| \to +\infty$ and therefore a sequence (λ_n) with the required properties exists. The Riesz product μ is then a singular measure ([6], V, 7.5 and 7.6). To prove the lemma it is sufficient to show that

$$\sum_k |\hat{\mu}(k)\, \hat{u}(k)| < +\infty \quad .$$

$\hat{\mu}(k) \neq 0$ if and only if

$$k = \alpha_1 \lambda_1 + \alpha_2 \lambda_2 + \ldots + \alpha_n \lambda_n$$

for some $n = 1,2,\ldots$, and $\alpha_i = -1,0,1$, and then

$$\hat{\mu}(k) = \frac{1}{2\pi} \left(\frac{1}{2}\right)^{|\alpha_1|+\ldots+|\alpha_n|} .$$

Let Γ_n be the set $\left\{k : k = \alpha_1 \lambda_1 + \ldots + \alpha_n \lambda_n ,\ \alpha_i = -1,0,1 ,\ i = 1,\ldots,n ,\ \text{and}\ |\alpha_n| = 1\right\}$.
Obviously

$$\sum_k |\hat{\mu}(k)\, \hat{u}(k)| = \frac{1}{2\pi} |\hat{u}(0)| + \sum_{n=1}^{+\infty} \sum_{k\in\Gamma_n} |\hat{\mu}(k)\, \hat{u}(k)| .$$

If $k \in \Gamma_n$ then

$$|k| \geq \lambda_n - (\lambda_1 + \ldots + \lambda_{n-1}) \geq \lambda_n/2$$

and consequently

$$\sum_k |\hat{\mu}(k)\, \hat{u}(k)|$$

$$\leq \frac{1}{2\pi}\,|\hat{u}(0)| + \sum_{n=1}^{+\infty} \Big(\sum_{k\in\Gamma_n} |\hat{\mu}(k)|\Big) \sup_{|k|\geq\lambda_n/2} |\hat{u}(k)|$$

$$\leq \frac{1}{2\pi}\,|\hat{u}(0)| + \frac{1}{2\pi} \sum_{n=1}^{+\infty} 2^{n-1}(2^{-n}\, n^{-2}) < +\infty$$

because

$$\sum_{k\in\Gamma_n} |\hat{\mu}(k)| = \frac{1}{2\pi}\, 2^{n-1}$$

and

$$\sup_{|k|\geq\lambda_n/2} |\hat{u}(k)| \leq 2^{-n}\, n^{-2} .$$

The analogous proof is valid for $d > 1$. The measure μ should be only defined as the Riesz product of the form

$$\prod_{n=1}^{+\infty} (1 + \cos \lambda_n x_1) \cdot \ldots \cdot (1 + \cos \lambda_n x_d) .$$

4. Proof of the Theorem 1. We can consider only the case $E = T^d$. To see this let a mapping $\gamma : R^d \to T^d$ be given by the formula

$$\gamma(x_1,\ldots,x_d) = (e^{ix_1}, e^{ix_2}, \ldots, e^{ix_d}) , \quad (x_1,\ldots,x_d) \in R^d .$$

Then the image $Y = \gamma(X)$ of the process X is also a p.w.i.i. (for the definition of the process Y we refer to [3] Theorem 10.25). A Borel set $B \subset T^d$ is polar (semipolar, null) for Y if the set $\gamma^{-1}(B)$ is polar (semipolar, null) for X. Obviously $\xi_d(B) = 0$ if and only if $\xi_d(\gamma^{-1}(B)) = 0$. Therefore in the sequel we shall assume $E = T^d$.

Instead of dealing with the process X it is more convenient to consider a new process (which we shall denote also by X) corresponding to the semigroup $(e^{-t}\,\mu_t)$ - the process obtained by "killing" X at the rate $1\,dt$.

Let us assume now that every compact set $K \subset T^d$ such that $\xi_d(K) = 0$ is semipolar for X. Then the measure ξ is a reference measure for X. Therefore there exists a lower semicontinuous excessive for X function u such that

$$Uf(x) \overset{df}{=} E^x \left(\int_0^{+\infty} f(X_s) ds \right) = \int_{T^d} u(x-y)\, f(y)\, \xi(dy) \quad ,$$

for any Borel function $f \geq 0$ on T^d ; since $U1 = \text{constant} < +\infty$ therefore $u \in L^1(T^d)$. By virtue of the Lemma there exists a singular probability measure μ on T^d such that the potential $U\mu$ of μ defined as

$$U\mu(x) = \int_{T^d} u(x-y)\, \mu(dy) = u * \mu(x) \ , \quad x \in T^d$$

is, after a modification, a continuous function. Let f be a such modification. Then

$$\beta \overline{U}^\beta f = \beta U^\beta U\mu$$

for all $\beta > 0$ and $\beta U^\beta U\mu \uparrow U\mu$ as $\beta \uparrow +\infty$ (see [1, p.72]). But $\beta \overline{U}^\beta(x,\cdot) \to \delta_{\{x\}}$ weakly as $\beta \to +\infty$ thus $f = U\mu$ and the potential $U\mu$ is a continuous function on the whole T^d . By virtue of [1,p.285, Prop.(4.3)] the measure μ charges no polar sets. To prove that, in fact, μ charges no semipolar sets assume that on the contrary, there exists a compact semipolar set $K \subset T^d$ such that $\mu(K) > 0$. Denote by ν and ν' the restrictions of the measure μ respectively to the set K and to the K^c , the complement of K . Since $U\nu$, $U\nu'$ are lower semicontinuous and $U\nu = U\mu - U\nu'$ therefore the potential $U\nu$ is a continuous function. We shall show that

$$P_K\, U\nu(\cdot) = U\nu(\cdot)$$

where, as usual,

$$P_K\, U\nu(x) = E^x \left\{ U\nu(X_{T_K}) \right\}$$

and

$$T_K = \inf \left\{ t > 0 \ , \ X_t \in K \right\}$$

Let (G_n) be a decreasing sequence of open sets $G_n \supset K$ such that $T_{G_n} \uparrow T_K$ almost surely P^ξ (see [1, p.62, Thm. 11.2]). Quasi - left - continuity of the process X implies that

$$X_{T_{G_n}} \to X_{T_K}$$

almost surely P^ξ. But the measure ν is concentrated on the open set G_n therefore for all $x \in T^d$:

$$E^x \left\{ U\nu(X_{T_{G_n}}) \right\} = P_{G_n} U\nu(x) = U\nu(x) \ .$$

From this and continuity of $U\nu$ we obtain that

$$U\nu = E \left\{ U\nu(X_{T_{G_n}}) \to E^{\cdot} \left\{ U\nu(X_{T_K}) \right\} \right.$$

almost surely ξ. Thus $U\nu = P_K U\nu$ ξ - almost surely and consequently $U\nu = P_K U\nu$ on the whole T^d .

Let $(A(t))$ be the natural potential associated with U (see [1], p. 281, Thm. 3.4 and p. 180, Thm. 4.22), and let $T_1 = T_K$, $T_{n+1} = T_n + T_K \cdot \theta_{T_n}$. We can assume that $T_n \uparrow +\infty$ ([1], p. 80, proof of Prop. 3.4). Obviously

$$P_K U\nu(x) = E^x \left\{ A(\infty) - A(T_K) \right\} = U\nu(x) = E^x \left\{ A(\infty) \right\} \ .$$

Therefore, for every $x \in T^d$, $E^x \left\{ A(T_K) \right\} = 0$. Analogously $E^x \left\{ A(T_n) \right\} = 0$ for $n = 1,2,\ldots$ and since $T_n \uparrow +\infty$ we obtain $E^{\cdot}\{A(\infty)\} = U\nu(\cdot) \equiv 0$, By virtue of [1], (p. 260, Prop. 1.15) $\nu \equiv 0$. This contradicts our assumption $\mu(K) = \nu(K) > 0$. To complete the proof of the Theorem 1 it is sufficient to remark that since the measure μ is singular and charges no semipolar sets therefore there exists a compact set K such that $\xi(K) = 0$ and K is not semipolar

COROLLARY. <u>For any p.w.i.i. X the family of all ξ - negligible Borel sets is essentially larger than the family of all polar Borel sets for X . This answers the question posed in</u> [5], p. 235.

THEOREM 2. <u>Let X be a p.w.i.i. If either there exists a reference measure for X or excessive functions of X are lower semicontinuous then the family of all null Borel sets for X is essentially larger than the family of all semipolar Borel sets for X</u> .

<u>Proof</u>. The assumptions of the theorem are equivalent to the statement: a Borel set $A \subset E$ is null if and only if $\xi(A) = 0$ (see e.g. [5], Thm. 1.4). Thus it is sufficient to apply Theorem 1.

<u>Remark</u>. If X is a Poisson process then family of semipolar sets for X consists of the empty set only, thus in this case trivially a set is null if and only if it is semipolar. This example shows that for validity of Theorem 2 some assumptions are necessary.

<u>Problem</u>. Generalize Theorem 2 for other types of standard Markov processes.

<u>Acknowledgment</u>. I would like to thank Professors J.P. Kahane and P.A. Meyer for discussions on the subject of this note.

<u>References</u>

[1] R.M.Blumenthal and R.K.Getoor, Markov Processes and Potential Theory, Academic Press, New York and London, 1968.

[2] C.Dellacherie, Une conjecture sur les ensembles semi-polaires, Lecture Notes in Mathematics 321, Seminaire de Probabilites VII, Springer-Verlag, 1973, pp. 51-57.

[3] E.B.Dynkin, Markov processes, Fizmatgiz, Mascow 1963.

[4] G.A.Hunt, Markoff processes and potentials, IJM 2(1958) pp. 151-213.

[5] J.Zabczyk, Sur la theorie semi-classique du potential pour les processes a accroissements independants, Studia Math., 35(1970) pp. 227-247.

[6] Zygmund, Trigonometric Series, Vol. 1, Cambridge, 1959.

Vol. 309: D. H. Sattinger, Topics in Stability and Bifurcation Theory. VI, 190 pages. 1973. DM 20,-

Vol. 310: B. Iversen, Generic Local Structure of the Morphisms in Commutative Algebra. IV, 108 pages 1973. DM 18,-

Vol. 311: Conference on Commutative Algebra. Edited by J. W. Brewer and E. A. Rutter. VII, 251 pages. 1973. DM 24,-

Vol. 312: Symposium on Ordinary Differential Equations. Edited by W. A. Harris, Jr. and Y. Sibuya. VIII, 204 pages. 1973. DM 22,-

Vol. 313: K. Jörgens and J. Weidmann, Spectral Properties of Hamiltonian Operators. III, 140 pages. 1973 DM 18,-

Vol. 314: M. Deuring, Lectures on the Theory of Algebraic Functions of One Variable. VI, 151 pages 1973. DM 16,-

Vol 315: K Bichteler, Integration Theory (with Special Attention to Vector Measures). VI, 357 pages 1973 DM 29,-

Vol. 316: Symposium on Non-Well-Posed Problems and Logarithmic Convexity. Edited by R. J. Knops. V, 176 pages. 1973. DM 20,-

Vol. 317: Séminaire Bourbaki - vol. 1971/72. Exposés 400-417. IV, 361 pages. 1973 DM 29,-

Vol. 318: Recent Advances in Topological Dynamics Edited by A. Beck. VIII, 285 pages. 1973. DM 27,-

Vol. 319: Conference on Group Theory. Edited by R W Gatterdam and K. W. Weston. V, 188 pages. 1973. DM 20,-

Vol. 320: Modular Functions of One Variable I Edited by W. Kuyk V, 195 pages. 1973. DM 20,-

Vol. 321: Séminaire de Probabilités VII. Edité par P A Meyer. VI, 322 pages. 1973 DM 29,-

Vol. 322: Nonlinear Problems in the Physical Sciences and Biology. Edited by I Stakgold, D D Joseph and D. H Sattinger. VIII, 357 pages. 1973 DM 29,-

Vol. 323: J. L. Lions, Perturbations Singulières dans les Problèmes aux Limites et en Contrôle Optimal XII, 645 pages. 1973 DM 46,-

Vol. 324: K. Kreith, Oscillation Theory. VI, 109 pages 1973. DM 18,-

Vol. 325: C.-C. Chou, La Transformation de Fourier Complexe et L'Equation de Convolution. IX, 137 pages. 1973. DM 18,-

Vol. 326: A. Robert, Elliptic Curves. VIII, 264 pages 1973. DM 24,-

Vol 327: E Matlis, One-Dimensional Cohen-Macaulay Rings. XII, 157 pages 1973. DM 20,-

Vol. 328: J. R. Büchi and D. Siefkes, The Monadic Second Order Theory of All Countable Ordinals. VI, 217 pages. 1973. DM 22,-

Vol. 329: W. Trebels, Multipliers for (C, α)-Bounded Fourier Expansions in Banach Spaces and Approximation Theory. VII, 103 pages. 1973. DM 18,-

Vol. 330: Proceedings of the Second Japan-USSR Symposium on Probability Theory. Edited by G Maruyama and Yu. V. Prokhorov VI, 550 pages. 1973. DM 40,-

Vol. 331: Summer School on Topological Vector Spaces. Edited by L. Waelbroeck. VI, 226 pages. 1973. DM 22,-

Vol. 332: Séminaire Pierre Lelong (Analyse) Année 1971-1972 V, 131 pages. 1973 DM 18,-

Vol. 333: Numerische, insbesondere approximationstheoretische Behandlung von Funktionalgleichungen. Herausgegeben von R. Ansorge und W. Törnig VI, 296 Seiten. 1973 DM 27,-

Vol. 334: F. Schweiger, The Metrical Theory of Jacobi-Perron Algorithm. V, 111 pages 1973. DM 18,-

Vol. 335: H. Huck, R Roitzsch, U. Simon, W. Vortisch, R Walden, B. Wegner und W. Wendland, Beweismethoden der Differentialgeometrie im Großen. IX, 159 Seiten. 1973 DM 20,-

Vol. 336: L'Analyse Harmonique dans le Domaine Complexe. Edité par E. J. Akutowicz. VIII, 169 pages. 1973. DM 20,-

Vol. 337: Cambridge Summer School in Mathematical Logic. Edited by A. R. D. Mathias and H. Rogers. IX, 660 pages. 1973. DM 46,-

Vol: 338: J. Lindenstrauss and L Tzafriri, Classical Banach Spaces X, 243 pages. 1973. DM 24,-

Vol. 339: G. Kempf, F. Knudsen, D. Mumford and B. Saint-Donat, Toroidal Embeddings I. VIII, 209 pages. 1973. DM 22,-

Vol. 340: Groupes de Monodromie en Géométrie Algébrique. (SGA 7 II). Par P. Deligne et N. Katz. X, 438 pages. 1973. DM 44,-

Vol. 341: Algebraic K-Theory I, Higher K-Theories. Edited by H. Bass. XV, 335 pages. 1973. DM 29,-

Vol. 342: Algebraic K-Theory II, "Classical" Algebraic K-Theory, and Connections with Arithmetic. Edited by H. Bass. XV, 527 pages. 1973. DM 40,-

Vol. 343: Algebraic K-Theory III, Hermitian K-Theory and Geometric Applications. Edited by H. Bass. XV, 572 pages. 1973. DM 40,-

Vol. 344: A. S. Troelstra (Editor), Metamathematical Investigation of Intuitionistic Arithmetic and Analysis. XVII, 485 pages. 1973. DM 38,-

Vol. 345: Proceedings of a Conference on Operator Theory. Edited by P A. Fillmore. VI, 228 pages. 1973. DM 22,-

Vol. 346: Fučík et al., Spectral Analysis of Nonlinear Operators. II, 287 pages. 1973 DM 26,-

Vol. 347: J. M. Boardman and R. M. Vogt, Homotopy Invariant Algebraic Structures on Topological Spaces. X, 257 pages. 1973. DM 24,-

Vol. 348: A. M. Mathai and R. K. Saxena, Generalized Hypergeometric Functions with Applications in Statistics and Physical Sciences. VII, 314 pages. 1973. DM 26,-

Vol 349: Modular Functions of One Variable II. Edited by W. Kuyk and P. Deligne V, 598 pages. 1973. DM 38,-

Vol. 350: Modular Functions of One Variable III. Edited by W. Kuyk and J.-P. Serre. V, 350 pages. 1973. DM 26,-

Vol 351: H. Tachikawa, Quasi-Frobenius Rings and Generalizations. XI, 172 pages. 1973. DM 20,-

Vol. 352: J. D Fay, Theta Functions on Riemann Surfaces. V, 137 pages. 1973 DM 18,-

Vol. 353: Proceedings of the Conference. on Orders, Group Rings and Related Topics. Organized by J. S. Hsia, M. L. Madan and T. G. Ralley. X, 224 pages. 1973 DM 22,-

Vol. 354: K. J. Devlin, Aspects of Constructibility. XII, 240 pages. 1973 DM 24,-

Vol. 355: M. Sion, A Theory of Semigroup Valued Measures. V, 140 pages. 1973. DM 18,-

Vol. 356: W. L. J van der Kallen, Infinitesimally Central-Extensions of Chevalley Groups. VII, 147 pages. 1973. DM 18,-

Vol 357: W Borho, P Gabriel und R. Rentschler, Primideale in Einhüllenden auflösbarer Lie-Algebren V, 182 Seiten. 1973. DM 20,-

Vol. 358: F. L. Williams, Tensor Products of Principal Series Representations. VI, 132 pages. 1973. DM 18,-

Vol. 359: U. Stammbach, Homology in Group Theory. VIII, 183 pages. 1973 DM 20,-

Vol. 360: W. J. Padgett and R. L. Taylor, Laws of Large Numbers for Normed Linear Spaces and Certain Fréchet Spaces. VI, 111 pages 1973. DM 18,-

Vol. 361: J. W. Schutz, Foundations of Special Relativity: Kinematic Axioms for Minkowski Space Time. XX, 314 pages. 1973. DM 26,-

Vol. 362: Proceedings of the Conference on Numerical Solution of Ordinary Differential Equations. Edited by D. Bettis. VIII, 490 pages. 1974. DM 34,-

Vol. 363: Conference on the Numerical Solution of Differential Equations. Edited by G. A. Watson. IX, 221 pages 1974 DM 20,-

Vol 364: Proceedings on Infinite Dimensional Holomorphy. Edited by T. L. Hayden and T. J. Suffridge. VII, 212 pages. 1974. DM 20,-

Vol. 365: R. P. Gilbert, Constructive Methods for Elliptic Equations. VII, 397 pages. 1974. DM 26,-

Vol 366: R. Steinberg, Conjugacy Classes in Algebraic Groups (Notes by V V. Deodhar). VI, 159 pages 1974. DM 18,-

Vol. 367: K. Langmann und W. Lütkebohmert, Cousinverteilungen und Fortsetzungssätze. VI, 151 Seiten. 1974. DM 16,-

Vol. 368: R J. Milgram, Unstable Homotopy from the Stable Point of View. V, 109 pages. 1974. DM 16,-

Vol. 369: Victoria Symposium on Nonstandard Analysis. Edited by A. Hurd and P. Loeb XVIII, 339 pages. 1974. DM 26,-